Hydrology for Engineers and Planners

A. T. HJELMFELT, JR.

J. J. CASSIDY

HYDROLOGY
FOR
ENGINEERS
AND
PLANNERS

IOWA STATE UNIVERSITY PRESS / AMES, IOWA

1975

ALLEN T. HJELMFELT, JR., Professor of Civil Engineering at the University of Missouri, Columbia, received the Ph.D. degree from Northwestern University. He is the author of several technical papers and reports in the fields of fluid mechanics and water resource engineering. He is a registered professional engineer and serves as a consulting engineer.

JOHN J. CASSIDY is Assistant Chief Hydraulic Engineer with the Bechtel Corporation, San Francisco, Calif., and formerly Professor and Chairman of Civil Engineering at the University of Missouri, Columbia. He received the Ph.D. degree from the University of Iowa and was a design engineer with the Montana State Water Conservation Board. He is a registered professional engineer and the author of several technical articles. He has conducted international short courses on hydrology and hydraulic engineering and has been active in research.

© 1975 The Iowa State University Press
Ames, Iowa 50010. All rights reserved

Composed and printed by
The Iowa State University Press

First edition, 1975

Library of Congress Cataloguing in Publication Data

Hjelmfelt, A   T   1937–
    Hydrology for engineers and planners.

    Includes bibliographical references.
    1. Hydrology.   2. Civil engineering.   I. Cassidy, John Joseph, 1930–
joint author.   II. Title.
GB661.H55   627   74–11091
ISBN 0–8138–0795–6

CONTENTS

Preface ix

1 SCOPE OF HYDROLOGY 3
   Objective of Hydrologic Analysis 4
   Hydrologic Cycle 5
   Response of a Watershed 8
   Hydrologic Data 9
   Problems 13
   References 14

2 STATISTICAL METHODS IN HYDROLOGY 15
   Elements of Probability 15
   Return Period 21
   Some Hydrologic Applications of Probability 21
   Elements of Statistics 23
   Normal Distribution 26
   Log Pearson Type III Distribution 29
   Graphic Methods 32
   Problems 38
   References 39

**3 APPLICATION OF PROBABILITY AND STATISTICS
   TO SELECTION OF A DESIGN DISCHARGE** 41
   Evaluation of Alternate Levee Designs 41
   Selection of Culvert Size 45

Problems 49
References 50

4 PRECIPITATION 51
Elements of Meteorology 51
Analysis of Precipitation at a Point 58
Analysis of Precipitation over an Area 71
Construction of Synthetic Storms 74
Probable Maximum Precipitation 77
Problems 83
References 84

5 INFILTRATION 85
Infiltration Process 85
Infiltration Indices 87
Estimation of Rainfall Excess by Soil-Cover Complex
            Analysis 88
Problems 95
References 95

6 STREAMFLOW 96
The Hydrograph 96
Baseflow Recession Curves 99
Complex Hydrographs 103
The Unit Hydrograph 103
Application of the Unit Hydrograph 106
Synthetic Unit Hydrographs 107
Equations for Determination of Peak Flow 113
Problems 116
References 118

7 FLOOD ROUTING 120
Reservoir Routing 121
Streamflow Routing 126
Baseflow 133
Problems 133
References 136

8 RESERVOIR YIELD **137**

Simulation Methods 137

Nonsequential Drought 141

Probabilistic Method 144

Problems 149

References 151

9 GROUNDWATER **152**

Water in the Zone of Aeration 152

Water in the Zone of Saturation 153

Darcy's Law 153

Two-dimensional Flows 156

Steady-State Well Flows 159

Unsteady-Flow Well Problems 160

Problems 166

References 167

10 EVAPORATION **169**

The Process of Evaporation 169

Estimation of Evaporation 171

Problems 179

References 180

Appendix A: General Statistical Data **183**

Appendix B: Blue River Streamflow Data **194**

Appendix C: Little Blue River Streamflow Data **201**

Appendix D: Precipitation Data, Northwestern Missouri, July 17–20, 1965 .... **204**

Index **207**

PREFACE

THE STUDY OF HYDROLOGY involves so many variables that it is difficult, if not impossible, to foresee that it will ever approach the status of an exact science. Nevertheless, the engineer is faced with a need to design bridges, flood control structures, dams, spillways, street and highway drainage structures, channels, and entire water supply systems. Without exception these designs require quantitative hydrologic analysis and the reasonable choice of design events. As a result even though the science of hydrology is far from being completely understood, a number of analytic methods have evolved that are more or less accepted by the engineering profession. In the design of any hydraulic engineering structure a design flow must always be chosen. For structures of moderate cost this has been established over a period of years as a result of experience. In major projects, however, the design flow must be chosen on the basis of combined hydrologic, hydraulic, and economic factors. This book is intended to provide tools for use in the analysis of such problems.

This volume has been written for use as a textbook in a one-semester undergraduate course in hydrology; therefore certain limitations had to be imposed. As a result, where many different methods have evolved in processes such as the construction of unit hydrographs, only certain commonly used methods are presented here. In particular, the book stresses the need to synthesize rainfall or streamflow where historical records are not available. Emphasis throughout is on the application of hydrologic principles, not upon the collection of hydrologic data since the collection of such data is generally limited to particular agencies. After having mastered the material in this text, a student should be capable of approaching a comprehensive hydrologic analysis as required in engineering projects.

The book evolved from the preparation of lecture notes, which have been used over a period of nearly five years in classes at the University of Missouri. As a result, student input and approach to the material has had a definite influence upon the final form and the problems presented. Because hydrology at the University of Missouri has been offered to sophomores in civil engineering, the textbook does not require a knowledge of fluid mechanics or the principles of hydraulics. Such material is introduced where it is needed.

As the text was used in note form, a valuable part of the course was found to center around a comprehensive term problem. Representative problems have included the analysis of actual reservoirs and spillways in associated flood control projects with regard to their capability of passing the design flood safely. Other projects included choice of the economically optimum height of levee for flood protection. Projects such as these require information on topography, stream characteristics, extreme flood events, and precipitation. Ideally, such information should be gathered by the student from library references or local agencies. However, for the sake of convenience some general data and data for selected drainage basins have been presented in the Appendices.

ALLEN T. HJELMFELT, JR.

JOHN J. CASSIDY

Hydrology for Engineers and Planners

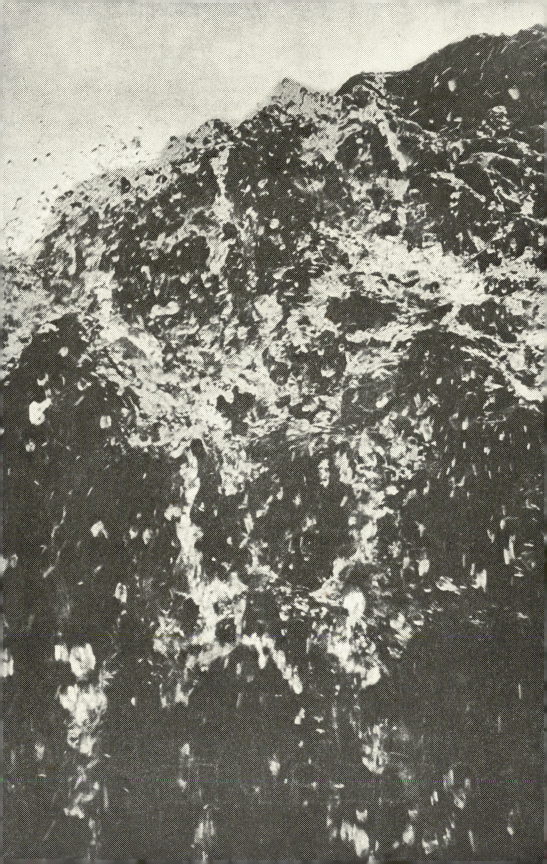

CHAPTER ONE
SCOPE OF HYDROLOGY

ESTIMATION OF SOME ASPECT of the quantity of water available is of prime importance in all water resource engineering. The analysis is basic to the planning, design, and operation of water resource systems. These systems may vary in size from a culvert on a country road to an integrated development of reservoirs, levees, and canals in a large river basin. Though the scope of consideration varies, the principles of analysis remain the same. A multipurpose project may include:

1. Water supply for municipalities and industries.
2. Flood plain management and flood damage reduction.
3. Hydropower for electric energy production.
4. Navigation.
5. Irrigation and drainage.
6. Watershed management for agricultural crop uses, including soil and water conservation and erosion control.
7. Water quality improvement for pollution control.
8. Water-oriented outdoor recreation.
9. Fish and wildlife propagation.

These require various approaches and considerations in the estimation of water quantity. Flood control primarily involves peak rates of discharge, whereas supply for most other uses involves analysis of the periods of lowest flow.

Hydrology is aimed at determining design parameters, which are analogous to design loads in structural analysis. The results are always estimates. Accuracy is limited in many cases, and sometimes only rough estimates can be obtained. These estimates are seldom less accurate than those of structural design loads or highway traffic volumes, however.

Analysis involving such uncertainties is usually accomplished with *probabilistic methods,* whereas *deterministic methods* are used where uncertainty is not admitted.

OBJECTIVE OF HYDROLOGIC ANALYSIS

The objective of a hydrologic analysis is best indicated by an example. One example can hardly indicate the scope of hydrology but does provide a definite basis for discussion.

A mobile home park is located in the flood plain of a small stream and is subject to frequent flooding. The rapidity with which the floods occur limits the possibility of evacuation or temporarily moving the mobile homes to higher ground.

An analysis of alternate means of coping with the problem should be investigated. These might include doing nothing, constructing a levee around the trailer park, or abandoning the current site and moving to higher ground. A hydrologic investigation is required to compare the various alternatives. For example, a very high levee would provide a high degree of protection, but the levee cost would be correspondingly large. A lower levee would provide less protection, but the levee cost would also be less. The aim is to balance the cost of protection with the expected cost of damages so that probable benefits justify the cost.

The estimated damage costs for various levels of flooding are given in Table 1.1, and the estimated annual costs of various sizes of levees are shown in Table 1.2. Table 1.3 indicates the stage elevation at the mobile home park for various flows in the channel. A map of the watershed is given in Figure 1.1.

This example could be solved by using the material of Chapters 2 and 3 if adequate streamflow records were available at the site, but they are not. Rainfall information is available, however, thus streamflow information must be estimated from rainfall records. The process of synthesizing streamflow information from rainfall measurements is given in Chapters 4, 5, and 6. Hence this example problem may be considered after Chapter 6 has been covered.

TABLE 1.1. Estimated flood damage

Flood stage elevation (ft)	Flood damage	Flood stage elevation (ft)	Flood damage
615	$ 0	619	$ 17,500
616	1,000	620	70,000
617	3,000	621	120,000
618	11,000	622	180,000

TABLE 1.2. Estimated annual levee costs (base of levee at elevation 615 ft)

Height of levee (ft)	Annual cost of structure	Annual cost of maintenance
1	$ 100	$100
2	163	100
3	284	150
4	592	150
5	808	200
6	1,184	250
7	1,720	300
8	2,416	350
9	3,272	400

A flood-control reservoir may also be considered as a means of providing flood protection. The modification of the flood wave by the reservoir is discussed in Chapter 7. If this reservoir is also to supply water for times of low flow, the dependable quantity of water must be determined. This is the subject of Chapter 8. Evaporation from the reservoir will result in a loss of available water, and this is covered in Chapter 10. The use of groundwater as an additional source of water is discussed in Chapter 9.

Since the transformation of rainfall to streamflow is basic to hydrologic analysis, a general description follows of the physical process called the hydrologic cycle.

HYDROLOGIC CYCLE

Water occurs in many places and in many phases on, in, and over the earth. The transformation from one phase to another and the motion from one location to another constitute the hydrologic cycle, which is a closed system having no beginning nor end. The cycle is depicted in Figure 1.2 (1).

Atmospheric moisture travels toward the earth's surface in the form of rain, hail, snow, or condensation. A portion of this will be retained on buildings, trees, shrubs, and plants. This water never reaches the ground, and the quantity so retained is called interception loss. The

TABLE 1.3. Stage-discharge relation at mobile home park

Stage elevation (ft)	Discharge (cfs)	Stage elevation (ft)	Discharge (cfs)
615	800	619	4,400
616	1,300	620	6,400
617	1,900	621	9,100
618	3,000	622	13,400

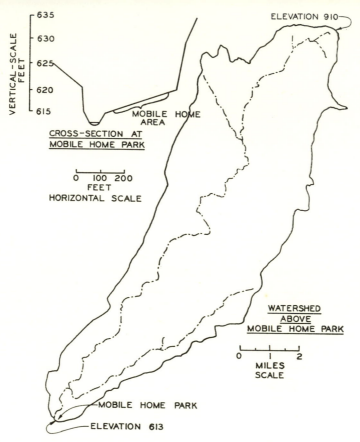

FIG. 1.1.   Watershed map and channel cross section of a levee project.

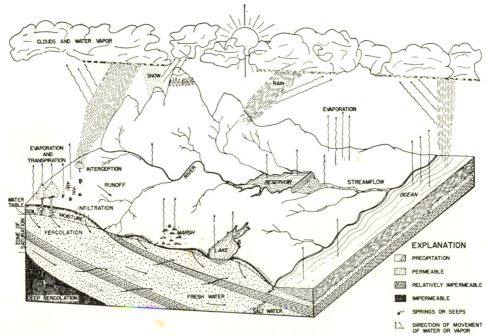

FIG. 1.2.   The hydrologic cycle (1).

water reaching the ground may follow several paths; some will evaporate back into the atmosphere, some will infiltrate into the earth. If the rainfall intensity exceeds the combined infiltration and evaporation rates, puddles will form. Water retained in puddles is said to be in depression storage.

As the puddles fill and overflow, water begins to move across the surface. The rainfall available for motion across the surface (water in excess of that evaporating, infiltrating, and held in depressions) is termed *rainfall excess*. Runoff cannot occur, however, until a layer of water covers the path of motion. The water contained in this path is said to be in detention storage. A portion of the runoff may infiltrate into the ground or may evaporate, returning to the atmosphere before reaching a stream or river.

The water that infiltrates into the ground first enters the soil zone which contains the roots of plants. This water may return to the atmosphere through evaporation from the soil surface or transpiration from plants. This upper soil zone can hold a limited quantity of water; this amount is known as the field capacity. If water is added to the zone when it is at field capacity, the water passes through to a lower zone (the zone of saturation or the groundwater zone). Water leaves the groundwater zone by capillary action into the root zone or by seepage into streams. Wells are drilled into the groundwater zone.

Figure 1.2 provides a schematic description of the hydrologic cycle, but it is not sufficiently idealized for the quantitative analysis needed in engineering. A system of tubes and reservoirs (a conceptual model) is used in Figure 1.3 (2) to represent the hydrologic cycle. The descrip-

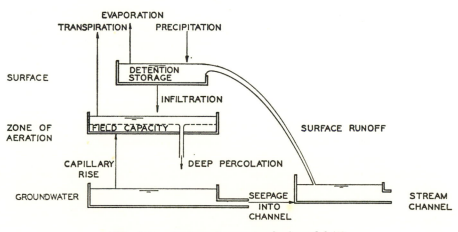

FIG. 1.3.   Engineering watershed model (2).

(1)	(2)	(3)	(4)	(5)
HYDROLOGIC PARAMETER	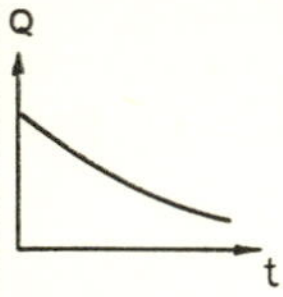 Q ⌐ t	 Q ⌐ t	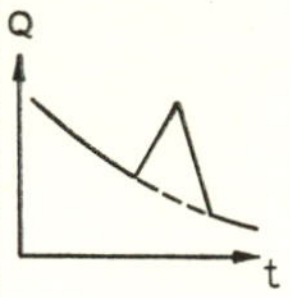 Q ⌐ t	 Q ⌐ t
RAINFALL INTENSITY	LESS THAN INFILTRATION RATE	LESS THAN INFILTRATION RATE	GREATER THAN INFILTRATION RATE	GREATER THAN INFILTRATION RATE
FIELD MOISTURE	TOTAL INFILTRATION LESS THAN FIELD CAPACITY	TOTAL INFILTRATION GREATER THAN FIELD CAPACITY	TOTAL INFILTRATION LESS THAN FIELD CAPACITY	TOTAL INFILTRATION GREATER THAN FIELD CAPACITY
SURFACE RUNOFF	NONE	NONE	YES	YES
FLOW INCREASE	NONE	GROUNDWATER ONLY	SURFACE RUNOFF ONLY	SURFACE RUNOFF AND GROUNDWATER

FIG. 1.4.   Classification of stream rises (3).

tion of Figure 1.2 applies equally well to Figure 1.3. This figure lends itself to mathematical modeling, whereas the hydrologic cycle as depicted in Figure 1.2 does not.

RESPONSE OF A WATERSHED

A watershed will respond differently to storms of different intensity and duration as well as to identical storms if the prior condition of the watershed varies. Some of the possibilities are indicated in Figure 1.4 (3), which may be explored in terms of the reservoir model of Figure 1.3.

Column 1 of Figure 1.4 indicates a number of characteristics of the land and the storm. The first row gives a series of streamflow hydrographs; the ordinate is discharge Q and the abcissa is time t.

Column 2 is headed by a hydrograph that is continuously decreasing. The flow results from seepage from the groundwater region to the stream. This type of hydrograph can result from no rainfall or from a storm whose intensity is less than the infiltration rate. In this case the water seeps into the ground as quickly as it falls so that no surface runoff occurs. In addition, the field capacity is not exceeded during the storm; thus no increase in groundwater storage results.

Column 3 is for conditions identical to those of column 2 except that the field capacity is exceeded during the storm, resulting in an increase in groundwater storage. The seepage into the stream increases with the storage, causing a slight rise in the streamflow.

In column 4 the rainfall intensity is greater than infiltration rate and surface runoff occurs. The field capacity is not exceeded; therefore, seepage from the groundwater zone does not increase.

In column 5 the rainfall intensity exceeds the infiltration rate, causing surface runoff. The field capacity is exceeded, creating an increase in groundwater that results in increased seepage into the stream.

HYDROLOGIC DATA

SOURCES OF DATA

In the chapters that follow emphasis will be on analysis of hydrologic data for engineering uses, not on the collection of the data. Much hydrologic and meteorologic data is collected, and summaries of the information are printed by various government agencies; however, these data are not always easily located.

The main source of meteorologic data is the Environmental Science Services Administration (ESSA). A general summary of meteorologic measurement is published monthly for each weather station. A monthly summary for each state or combination of states is published as "Climatological Data." More detailed precipitation data are published for each state or combination of states in "Hourly Precipitation Data." In addition, many publications are available giving summaries and compilations of specific meteorologic data. This material is most easily located through the ESSA publication, "Selective Guide to Published Climatic Data Sources" (4).

Streamflow data are gathered by many state and federal agencies. The most readily available source of information is published by the U.S. Geological Survey whose procedure is to print an annual volume for each state and a regional summary every five years. The regions are indicated on Figure 1.5, and the water supply papers containing the 1961–65 summaries are listed in Table 1.4 (5). The easiest way to locate the desired water supply papers is to obtain the most recent publication, which will also list earlier records. In addition, many of the state agencies publish summaries and analyses of hydrologic data, examples of which are "Low Flows of Missouri" and "Magnitude and Frequency of Missouri Floods," published by the agency, Missouri Geological Survey

TABLE 1.4. U.S. Geological Survey reports on stream discharges in the United States, 1961–65

Publication	Station numbers included in volume*		
1901. Part 1, North Atlantic slope basins—Vol. 1, Basins from Maine to Connecticut. 1,027 pages. 1969.	1–0100	to	1–2121
1902. Part 1, North Atlantic slope basins—Vol. 2, Basins from New York to Delaware. 924 pages. 1970.	1–3000	to	1–4845.5
1903. Part 1, North Atlantic slope basins—Vol. 3, Basins from Maryland to York River. 850 pages. 1970.	1–4848	to	1–6745
1904. Part 2, South Atlantic slope and eastern Gulf of Mexico basins—Vol. 1, Basins from James River to Savannah River. 942 pages. 1970.	2–0115	to	2–1985
1905. Part 2, South Atlantic slope and eastern Gulf of Mexico basins—Vol. 2, Basins from Ogeechee River to Carrabelle River. 772 pages. 1970.	2–1997	to	2–3304
1906. Part 2, South Atlantic slope and eastern Gulf of Mexico basins—Vol. 3, Basins from Apalachicola River to Pearl River. 774 pages. 1970.	2–3310	to	2–4926
1907. Part 3, Ohio River basin—Vol. 1, Ohio River basin above Kanawha River. 621 pages. 1971.	3–0080	to	3–1600.5
1908. Part 3, Ohio River basin—Vol. 2, Ohio River basin from Kanawha River to Louisville, Ky. 581 pages. 1971.	3–1601.1	to	3–2945
1909. Part 3, Ohio River basin—Vol. 3, Ohio River basin from Louisville, Ky., to Wabash River. 554 pages. 1971.	3–2945	to	3–3816
1910. Part 3, Ohio River basin—Vol. 4, Ohio River basin below Wabash River. 738 pages. 1971.	3–3820.25	to	3–6140
1911. Part 4, St. Lawrence River basin—Vol. 1, Basins of streams tributary to Lakes Superior, Michigan, and Huron.			
1912. Part 4, St. Lawrence River basin—Vol. 2, St. Lawrence River basin below Lake Huron.			
1913. Part 5, Hudson Bay and upper Mississippi River basins—Vol. 1, Hudson Bay basin.			
1914. Part 5, Hudson Bay and upper Mississippi River basins—Vol. 2, Upper Mississippi River basin above Keokuk, Iowa.			
1915. Part 5, Hudson Bay and upper Mississippi River basins—Vol. 3, Upper Mississippi River basin below Keokuk, Iowa.			
1916. Part 6, Missouri River basin—Vol. 1, Missouri River basin above Williston, N. Dak. 800 pages. 1969.	6–0110	to	6–3300
1917. Part 6, Missouri River basin—Vol. 2, Missouri River basin from Williston, N. Dak. to Sioux City, Iowa. 560 pages. 1969.	6–3300	to	6–4860
1918. Part 6, Missouri River basin—Vol. 3, Missouri River basin from Sioux City, Iowa, to Nebraska City, Nebr. 751 pages. 1969.	6–4860	to	6–8070
1919. Part 6, Missouri River basin—Vol. 4, Missouri River basin below Nebraska City, Nebr. 805 pages. 1969.	6–8070	to	6–9365
1920. Part 7, Lower Mississippi River basin—Vol. 1, Lower Mississippi River basin except Arkansas River basin. 1,103 pages. 1969.	7–0100	to	7–0782.1
	7–2660	to	7–3869.5

TABLE 1.4. (continued)

Year	Part	Description	From		To
1921.	Part 7,	Lower Mississippi River basin—Vol. 2, Arkansas River basin. 878 pages. 1969.	7–0820	to	7–2650
1922.	Part 8,	Western Gulf of Mexico basins—Vol. 1, Basins from Mermentau River to Colorado River. 967 pages. 1969.	8–0099.5	to	8–1625
1923.	Part 8,	Western Gulf of Mexico basins—Vol. 2, Basins from Lavaca River to Rio Grande. 786 pages. 1970.	8–1635	to	8–4925
1924.	Part 9,	Colorado River basin—Vol. 1, Colorado River basin above Green River. 488 pages. 1970.	9–0105	to	9–1870
1925.	Part 9,	Colorado River basin—Vol. 2, Colorado River basin from Green River to Compact Point. 618 pages. 1970.	9–1885	to	9–3830
1926.	Part 9,	Colorado River basin—Vol. 3, Lower Colorado River basin. 571 pages. 1970.	9–3830	to	9–5375
1927.	Part 10,	The Great Basin. 978 pages. 1970.	10–0100	to	10–4071.5
1928.	Part 11,	Pacific slope basins in California—Vol. 1, Basins from Tia Juana River to Santa Maria River. 501 pages. 1970.	11–0109	to	11–1410
1929.	Part 11,	Pacific slope basins in California—Vol. 2, Basins from Arroyo Grande to Oregon state line except Central Valley. 673 pages. 1970.	11–1413 11–4569.5	to to	11–1851.5 11–5330
1930.	Part 11,	Pacific slope basins in California—Vol. 3, Southern Central Valley basins. 655 pages. 1970.	11–1853	to	11–3375
1931.	Part 11,	Pacific slope basins in California—Vol. 4, Northern Central Valley basins. 611 pages. 1970.	11–3395	to	11–4550
1932.	Part 12,	Pacific slope basins in Washington—Vol. 1, Pacific slope basins in Washington except Columbia River basin. 679 pages. 1971.	12–0095	to	12–2151
1933.	Part 12,	Pacific slope basins in Washington—Vol. 2, Upper Columbia River basin. 685 pages. 1971.	12–3000	to	12–5140
1934.	Part 13,	Snake River basin. 776 pages. 1971.	13–0105	to	13–3530.5
1935.	Part 14,	Pacific slope basins in Oregon and lower Columbia River basin. 1957 pages. 1971.	14–0100	to	14–3789
1936.	Part 15,	Alaska.			
1937.	Part 16,	Hawaii and other Pacific areas.			

Source: *Water Resources Review* (5).

Note: The U.S. Geological Survey has published a 5-year compilation of records of streamflow and of the elevation and contents of lakes and reservoirs in the United States during the period October 1, 1960, through September 30, 1965. Daily streamflows are given as well as monthly and annual flow summaries. Each report is identified by a number in the water supply paper series of publications as well as by a part (region) and volume (subregion) number. The part and volume numbers of each report correspond to those shown on the map of Figure 1.5, thus identifying the geographical area covered by the data in the report. Many state, municipal, and private organizations have assisted by furnishing data or helping to collect data.

*The numbers as listed are shown in the short form formerly in use. The full numbers, designed for computer purposes and printouts, consist of 8 digits with no punctuation. For example, the full numbers for 1–4845.5 and 14–3789 are 01484550 and 14378900, respectively.

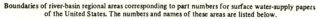

Boundaries of river-basin regional areas corresponding to part numbers for surface water-supply papers of the United States. The numbers and names of these areas are listed below.

Part number	Region	Part number	Region
1.	North Atlantic slope basins (St. John River to York River).	6.	Missouri River basin.
1–A.	Maine to Connecticut	6–A.	Missouri River basin above Sioux City, Iowa.
1–B.	New York to York River	6–B.	Missouri River basin below Sioux City, Iowa.
2.	South Atlantic slope and eastern Gulf of Mexico basins (James River to Pearl River).	7.	Lower Mississippi River basin.
		8.	Western Gulf of Mexico basins.
2–A.	James River to Savannah River.	9.	Colorado River basin.
2–B.	Ogeechee River to Pearl River.	10.	The Great Basin.
3.	Ohio River basin.	11.	Pacific slope basins in California.
3–A.	Ohio River basin except Cumberland and Tennessee River basins.	12.	(Prior to 1914:) North Pacific drainage basins.
3–B.	Cumberland and Tennessee River basins.	12.	(1914 to present:) Pacific slope basins in Washington and upper Columbia River basin.
4.	St. Lawrence River basin.	13.	Snake River basin.
5.	Hudson Bay and upper Mississippi River basins.	14.	Pacific slope basins in Oregon and lower Columbia River basin.

FIG. 1.5.   Boundaries of river basin regional areas and corresponding part numbers of surface water supply papers (5).

and Water Resources. Examples of summarized streamflow data and other pertinent hydrologic data are given in the Appendices.

COLLECTION OF DATA

The collection of hydrologic data is not covered in this book. Should the reader, by interest or necessity, desire to pursue the measurement process, references (6), (7), and (8) are recommended.

CONCLUSION

In an elementary text it is impossible to give a complete treatment of hydrology. It is hoped that a foundation for further development can be gained by pursuing the material presented. The most comprehensive reference on hydrology (in both scope and depth) is *Handbook of Applied Hydrology*, edited by Chow (9).

Technical journals such as *Water Resources Research* of the American Geophysical Union, the *Journal of the Hydraulics Division* and *Journal of the Irrigation and Drainage Division* of the American Society of Civil Engineers, and the *Water Resources Bulletin* of the American Water Resources Association regularly carry papers dealing with problems in hydrology.

PROBLEMS

1.1. a. *Exactly* what conditions must prevail if a rainstorm occurring on a watershed does not produce a rise in the hydrograph of the stream that drains the watershed?
   b. *Exactly* what conditions must prevail if a storm produces a rise in the streamflow hydrograph but does not produce any surface runoff?

1.2. Given the rainfall and corresponding streamflow hydrograph shown in Figure P1.2, classify each storm according to the following:
   a. Is the rainfall intensity more or less than the infiltration rate?
   b. Is there any increase in water in the groundwater storage zone?

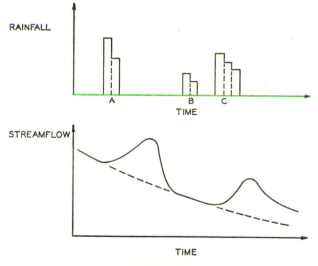

FIG. P1.2

1.3. Make the following conversions:

$$
\begin{array}{r}
8000 \text{ cubic feet} = \underline{\hspace{2cm}} \text{acre-inches} \\
200 \text{ cfs} = \underline{\hspace{2cm}} \text{acre-inches/hour} \\
200 \text{ cfs} = \underline{\hspace{2cm}} \text{square mile-inches/hour} \\
200 \text{ second-feet-days} = \underline{\hspace{2cm}} \text{acre-feet} \\
100 \text{ acre-inches/hour} = \underline{\hspace{2cm}} \text{square mile-inches/hour}
\end{array}
$$

Note: A second-foot-day is the volume obtained using a flow rate of one cfs continuously for one day. A cfs is sometimes called a cusec.

1.4. Obtain the record of annual peak discharges and mean annual flows for a river in your vicinity.

REFERENCES

1. Jones, P. B.; Walker, G. D.; Harden, R. W.; and McDaniels, L. L. The development of the science of hydrology. Texas Water Comm. Circ. 63-03, 1963.
2. Dowdy, D. R.; and O'Donnel, T. Mathematical models of catchment behavior. *ASCE J. Hydraul. Div.*, Vol. 91, Hy 4, 1965.
3. Horton, R. E. Surface runoff phenomena. 1. Analysis of the hydrograph. Horton Hydrological Laboratory, Voorheesville, N.Y., 1935.
4. U.S. Environmental Science Services Administration. Selective guide to published climatic data sources. *In* Key of Meteorological Records Documentation, No. 4.11, Washington, D.C., 1963.
5. *Water Res. Rev.*, Sept. 1971, Feb. 1972.
6. U.S. Weather Bureau. Observing Handbook 2. Substation Observations, Washington, D.C., 1970.
7. U.S. Geological Survey. *Techniques of Water-Resources Investigations,* 1969, 1970. (Series of books, each chapter published separately, detailing methods of collection and analysis of surface water, groundwater, sedimentation, and water quality data.)
8. U.S. Dept. of Agriculture. Field manual for research in agricultural hydrology. Agr. Handbook 224, Washington, D.C., 1966.
9. Chow, Ven Te, ed. *Handbook of Applied Hydrology.* McGraw-Hill, 1964.

CHAPTER TWO

STATISTICAL METHODS IN HYDROLOGY

NATURAL OCCURRENCES such as rain, floods, snow, and temperature vary in a seasonal fashion. In the Northern Hemisphere the January temperatures average considerably lower than do those in July. In other areas of the world, such as northern India, rainfall occurs almost entirely during the monsoon season, and virtually no rain falls during the remainder of the year.

Extremes in these natural occurrences, their annual averages, and almost any other measure of magnitude vary in a way that may be considered random. If precipitation is plotted against time as in Figure 2.1, it is clear that an average precipitation exists and that any individual annual precipitation deviates from this. Because of its seemingly erratic behavior, for the purposes of analysis the annual precipitation is considered to be a random variable. Annual maximum rainfalls and maximum wind speeds are further examples of natural events that are frequently assumed to occur in a random manner.

Techniques of probability and statistics are employed in the analysis of random events. In the following sections some elementary concepts of probability and statistics are presented. Rigorous derivations have been omitted, but basic elements are indicated by demonstration.

ELEMENTS OF PROBABILITY

In engineering design, conditions that must be accommodated are not always determinant. A tall building must be designed to resist wind loads. Highway drainage structures (storm drainage systems,

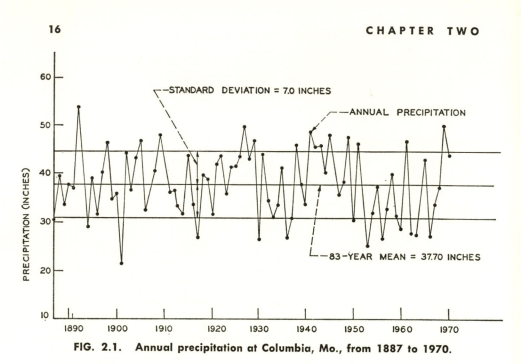

FIG. 2.1.   Annual precipitation at Columbia, Mo., from 1887 to 1970.

bridges, culverts) must be designed to safely pass a flood of some size. What magnitude of the design event should be used by the concerned engineer? No matter how great the severity of the storm used as the basis for design, some probability always exists that the design event will be exceeded. If the useful life of the structure is to be M years, there will always be some risk that failure will occur within this period. What is the probability that the design event will be exceeded during the intended useful life of the project? The purpose of this section is to present an analysis by which that question can be answered.

Values of probability range from 0 to 1. An event has 0 probability of occurrence if it cannot possibly happen and a probability of 1 if it certainly will happen. A very simple physical example can be used to demonstrate the concept of probability. Consider a disk (Fig. 2.2) that

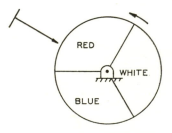

FIG. 2.2.   Game of chance.

can be spun about its center. The disk is divided into three equal pie-shaped sectors painted red, white, and blue respectively. When the disk is spun, it will eventually come to rest with a pointer directed at one of the three sectors. If the wheel is honest, what is the probability of obtaining red in one try? Since three outcomes are possible and they are each equally likely to occur, the probability of obtaining red on the first try is 1/3. What is the probability of obtaining either red or blue (we do not care which) on the first try? The area covered on the disk by the red and blue portions is 2/3 of the total area, thus the probability of obtaining either red or blue is 2/3. This value is the sum of the probabilities of the individual events; i.e., 2/3 equals the probability of getting a red plus the probability of getting a blue.

In general: The probability of obtaining either outcome A or outcome B, with A and B independent, is equal to the probability of obtaining outcome A plus the probability of obtaining outcome B.

$$P(A \text{ or } B) = P(A) + P(B) \qquad (2.1)$$

Another way of saying this, for a random experiment, is that the probability of obtaining any one of several independent outcomes is equal to the sum of the probabilities of each of the individual outcomes.

The independent qualification in the general statement warrants some comment. The disk in the example is divided into three distinct portions; thus all possibilities are independent. If, however, red and blue overlapped on a portion of the disk so that the stopped pointer indicated red and blue simultaneously, the red and blue outcomes would not be independent. The general statement would have to be modified to avoid counting the overlapping area twice.

Now suppose that the disk is spun twice. What is the probability of obtaining two reds in succession? Nine outcomes are possible from this experiment (see Fig. 2.3). Two reds in succession is one possibility (only one of nine); thus the probability of this sequence of outcomes is 1/9. This is the product of the probability of getting a red on the first spin and the probability of getting a red on the second spin.

In general: The probability of obtaining *both* outcome A and outcome B with A and B independent is the product of the probability of obtaining outcome A and the probability of obtaining outcome B.

$$P(A \text{ and } B) = P(A) \times P(B) \qquad (2.2)$$

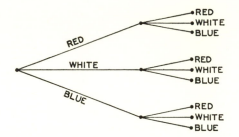

FIG. 2.3.   Possible combinations for two successive spins of the device of Figure 2.2.

For a sequence of independent random experiments this may also be stated as: The probability of obtaining any sequence of outcomes is equal to the product of the probabilities of each individual outcome.

In order to extend the concepts of probability toward more complex situations, consider the possible outcomes of tossing eight dice. What is the probability of obtaining exactly three sixes in one toss? Six equally likely outcomes are possible on each of the dice. Thus the probability of obtaining a six on a single die is 1/6, and the probability of obtaining something else is 5/6. An acceptable way of attaining the desired result would be to roll three sixes and then five nonsixes. The probability of rolling a six *and* a six *and* a six *and* five nonsixes is given by the product of the individual probabilities.

$$1/6 \cdot 1/6 \cdot 1/6 \cdot 5/6 \cdot 5/6 \cdot 5/6 \cdot 5/6 \cdot 5/6 = (1/6)^3 \cdot (5/6)^5 \quad (2.3)$$

This is the probability of one particular sequence. We do not care about the sequence, however, so consider throwing two nonsixes, three sixes, and three nonsixes. The probability of this particular sequence of outcomes is

$$5/6 \cdot 5/6 \cdot 1/6 \cdot 1/6 \cdot 1/6 \cdot 5/6 \cdot 5/6 \cdot 5/6 = (1/6)^3 \cdot (5/6)^5 \quad (2.4)$$

Thus, regardless of the order in which the sixes occur, the probability of obtaining exactly three sixes is the same.

$$(1/6)^3 \cdot (5/6)^5 \quad (2.5)$$

The requirement of three sixes would be met by the first sequence, *or* the second sequence, *or* any of the other sequences containing three sixes. Therefore, according to our rules, the probability of obtaining *exactly* three sixes when tossing eight dice is equal to the sums of the individual probabilities of the individual combinations.

$$P(3 \text{ times in 8 dice}) = N[(1/6)^3 \cdot (5/6)^5] \tag{2.6}$$

N is the number of possible combinations.

The number of possible combinations could be obtained by listing all the combinations. By digressing briefly, however, a general formula for N may be found, after which a general form of Eq. (2.6) will be given. Consider a bowl containing five beads labeled A, B, C, D, and E. How many different selections of three beads can be made?

While blindfolded (being able to see would eliminate chance in our selection) draw one bead. This first bead could be any of five; thus there are five different possibilities. Now draw the second bead. Since only four remain after the first draw, only four different possibilities exist for this second draw. Similarly, when the third bead is drawn, there are only three different possibilities. Thus the total number of possibilities for drawing three beads (permutations of five beads taken three at a time) is

$$5 \cdot 4 \cdot 3 = 5!/2! \tag{2.7}$$

This process is illustrated graphically in Figure 2.4.

Now imagine that the process of drawing three beads is repeated three times and the beads, in the order they were drawn, are ACB, CBA, and BAC. Actually, the same three beads are involved in each drawing and only the order of drawing is different. In order to determine the number of different obtainable combinations of beads, the number of drawing possibilities $5!/2!$ must be divided by the number of each combination that can be arranged. The arrangements of A, B, and C are ABC, ACB, BAC, BCA, CAB, CBA, or six ($3!$). Thus the total number of different combinations obtainable in drawing three beads from a bowl of five beads is

$$\frac{5!}{3! \cdot 2!} = \frac{5!}{3!(5-3)!} \tag{2.8}$$

In general: The number of different possible combinations of k things taken from a population of n things is $\binom{n}{k}$ given by

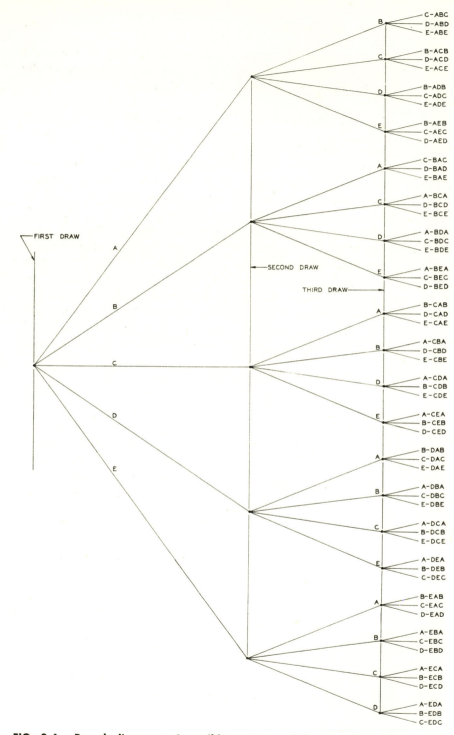

FIG. 2.4. Branch diagrams of possible outcomes of drawing three things from a total of five.

$$\frac{n!}{k!(n-k)!} = \binom{n}{k} \tag{2.9}$$

Returning now to the problem of throwing three sixes, the unknown quantity in Eq. (2.4) was the number of combinations of three sixes in eight dice, N. From Eq. (2.9)

$$N = \frac{n!}{k!(n-k)!} = \frac{8!}{3!(8-3)!} \tag{2.10}$$

so

$$P(3 \text{ sixes in } 8 \text{ dice}) = \frac{8!}{3!(8-3)!} (1/6)^3 (5/6)^5 = 0.104 \tag{2.11}$$

or in general:

$$P(k \text{ successes in } n \text{ trials}) = \binom{n}{k} p^k (1-p)^{n-k} \tag{2.12}$$

where p is the probability of success in any one trial.

RETURN PERIOD

In hydrology the reciprocal of the probability is frequently used. This quantity is termed the *return period* or sometimes *recurrence interval*. Using the symbol T_r for return period,

$$p = 1/T_r \tag{2.13}$$

Thus a flood that appears to have been exceeded on the average once in 20 years has a probability of exceedance in any one year of 1/20 or 0.05. Another way of stating this is that an annual peak flood that is exceeded on the average of one year in 20 years has a percent probability of $0.05 \times 100 = 5\%$ of being exceeded any year. Similarly, a flood having a return period of 50 years has a probability of 0.02 (2%) of being exceeded any year. It most certainly *does not* mean that every 50 years a flood of that magnitude will occur.

SOME HYDROLOGIC APPLICATIONS OF PROBABILITY

Equation (2.12) can be used to answer some questions regarding probability in hydrology:

1. What is the probability P that exactly three 50-year floods will occur in a single 100-year period?

$$P(3 \text{ in } 100) = \binom{100}{3} (1/50)^3 (49/50)^{97} = 0.183 \qquad (2.14)$$

There is an 18.3% chance that exactly three 50-year floods will occur in 100 years.

2. A more important question might be, What is the probability that three or more 50-year floods will occur in 100 years?

$$P(3 \text{ or more in } 100) = P(3 \text{ in } 100) + P(4 \text{ in } 100) + \ldots$$

Since the sum of the probabilities of all possible numbers of 50-year floods occurring in 100 years must equal unity, an easier way to answer our second question is

$$P(3 \text{ or more in } 100) = 1 - P(0 \text{ in } 100) - P(1 \text{ in } 100) - P(2 \text{ in } 100)$$

or

$$P(3 \text{ or more in } 100) = 1 - \binom{100}{0} (1/50)^0 (49/50)^{100}$$
$$- \binom{100}{1} (1/50)^1 (49/50)^{99} -$$
$$- \binom{100}{2} (1/50)^2 (49/50)^{98} \qquad (2.15)$$
$$= 1 - 0.133 - 0.271 - 0.273 = 0.323$$

Thus if an engineer designs a bridge to be safe during a 50-year flood, there is a 32.3% chance that the design flood will be exceeded three or more times in 100 years

3. In design the question of real interest is, What is the probability of having one or more failures in n years?

$$P(1 \text{ or more in } n \text{ years}) = 1 - P(0 \text{ in } n \text{ years})$$

or

$$P(1 \text{ or more in } n \text{ years}) = 1 - \binom{n}{0} (p)^0 (1 - p)^n$$

Thus the answer to Question 3 can be computed as:

$$P(\text{failure within } n \text{ years}) = 1 - (1 - p)^n \qquad (2.16)$$

Equation (2.16) was used to calculate the values in Table 2.1. According to this table if the designer will accept only a 1% chance of a flood control levee being overtopped in 25 years, he must design the levee to contain the 2440-year flood (this might well have been Noah's plight).

TABLE 2.1. Design return period as a function of project life and risk of failure

Permissible risk of failure (P)	Project life in years (n)			
	1	25	50	100
	Required return period $(1/p) = T_r$ (years)			
0.01	100	2,440	5,260	9,100
0.25	4	87	175	345
0.50	2	37	72	145
0.75	1.3	18	37	72
0.99	1.01	6	11	27

In constructing Table 2.1, only the concepts of probability and statistics were used. Thus the table is equally applicable to wind loads, snow loads, or any other random occurrence. To use Table 2.1, we must be able to relate a particular discharge to a given return period T_r; this is considered in the next sections of this chapter. Establishment of the permissible risk of failure is discussed in Chapter 3.

ELEMENTS OF STATISTICS

The role of statistics is to extract meaningful information from a mass of data. Records of precipitation amounts and flow rates of rivers kept by the U.S. Weather Bureau (1) and the U.S. Geological Survey (2, 3) are examples of masses of data that are difficult to use unless refined by some type of analysis. For example, we would like to determine the frequency with which a particular flow will be exceeded, by using a list of annual peak flows.

Consider the annual mean discharges of the Little Blue River (Appendix C). This information can be grouped as shown in Table 2.2 or displayed graphically as in Figure 2.5, which is called a histogram. We may also look at the annual mean discharges in cumulative terms as indicated in Table 2.3 and Figure 2.6.

The data used can be considered to be a small portion of all the annual mean flows that have occurred and will occur. If a much longer record were available, smaller group sizes could be used and smooth

TABLE 2.2. Numbers of mean annual discharges within ranges, Little Blue River

Discharge interval (cfs)	0–40	40–80	80–120	120–160	160–200	200–240	240–280
Number of occurrences	3	5	4	3	4	1	3

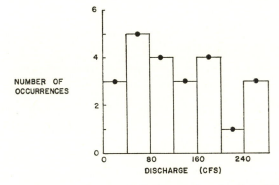

FIG. 2.5. Histogram of annual discharges for the Little Blue River.

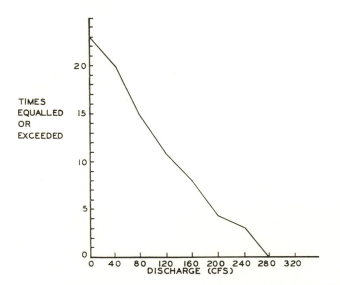

FIG. 2.6. Cumulative distribution of mean annual flows of the Little Blue River.

TABLE 2.3. Cumulative distribution of mean annual flows, Little Blue River

Discharge (cfs)	0	40	80	120	160	200	240	280
Number of times equalled or exceeded	23	20	15	11	8	4	3	0

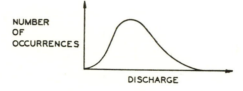

FIG. 2.7. Distribution of flows from a very long record.

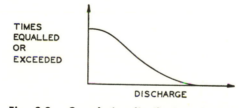

Fig. 2.8. Cumulative distribution of flows.

curves drawn. These might appear as in Figures 2.7 and 2.8. The record available constitutes a portion of the data that defines a smooth curve. If the exact form of the curve were known, the curve could be discussed instead of merely the mass of numbers collected from the records.

Usually we select a family of curves that fits the type of data quite closely and then try to define the particular curve in the family. For example, if a straight line fits most data of this type, we would try to establish the slope and intercept that define the particular line from the family of all straight lines.

The curves used in statistics are usually established by the mean, standard deviation, and skew. These parameters are defined by the following equations. Let

$$x_i = \text{magnitude of event } i$$
$$n = \text{total number of events}$$

then

$$\text{Mean} = \text{average} = M = \frac{\displaystyle\sum_{i=1}^{n} x_i}{n} \tag{2.17}$$

The mean locates the centroid of the area under the histogram.

$$\text{Standard deviation} = S = \sqrt{\frac{\displaystyle\sum_{i=1}^{n}(x_i - M)^2}{n-1}}$$

$$= \sqrt{\frac{\displaystyle\sum_{i=1}^{n}(x_i^2) - \left(\displaystyle\sum_{i=1}^{n} x_i\right)^2 \Big/ n}{n-1}} \tag{2.18}$$

The standard deviation is equivalent to the radius of gyration of the area under the histogram; therefore, it is a measure of the degree to which the area is spread away from the mean.

$$\text{Skew} = g = \left[\frac{\displaystyle\sum_{i=1}^{n}(x_i - M)^3}{n}\right]^{1/3}$$

$$= \left[\frac{n^2\left(\displaystyle\sum_{i=1}^{n} x_i^3\right) - 3n\displaystyle\sum_{i=1}^{n}(x_i)\displaystyle\sum_{i=1}^{n}(x_i^2) + 2\left(\displaystyle\sum_{i=1}^{n} x_i\right)^3}{n(n-1)(n-2)S^3}\right]^{1/3} \tag{2.19}$$

Other parameters of interest are

Median = half the magnitudes above and half below
Mode = most frequent magnitude

NORMAL DISTRIBUTION

In hydrology a wide variety of families of curves representing histograms have been tried. The family of curves with which most people

TABLE 2.4. Values of the normal distribution

Exceedance probability	K	Exceedance probability	K
0.0001	3.719	0.500	0.000
0.0005	3.291	0.550	—0.126
0.001	3.090	0.600	—0.253
0.005	2.576	0.650	—0.385
0.010	2.326		
0.025	1.960	0.700	—0.524
0.050	1.645	0.750	—0.674
		0.800	—0.842
0.100	1.282	0.850	—1.036
0.150	1.036	0.900	—1.282
0.200	0.842		
0.250	0.674	0.950	—1.645
0.300	0.524	0.975	—1.960
		0.990	—2.326
0.350	0.385	0.995	—2.576
0.400	0.253	0.999	—3.090
0.450	0.126	0.9995	—3.291
0.500	0.000	0.9999	—3.719

have some familiarity is the normal distribution. This is given by the equation

$$f = [1/(S\sqrt{2\pi})]\exp[- (x - M)^2/(2S^2)] \qquad (2.20)$$

This curve is defined by only two parameters, the standard deviation and the mean. Some properties of this distribution are that

$$f \rightarrow 0 \text{ as } x \rightarrow \pm\infty$$
$$f = \text{max at } x = M$$
Curve is symmetric about M

Usually we use a cumulative distribution, the integral of Eq. (2.20). The numerical value of the integral can be found using Table 2.4.

The normal distribution is sometimes used to describe mean annual discharges of a stream; however, it has the disadvantage of giving a probability to negative discharges. It seldom provides a good fit for flood discharges. To use the normal distribution to describe the mean annual discharge of a stream, the mean and standard deviation of the mean annual discharges are determined from the recorded values using Eq. (2.17) and Eq. (2.18). Probabilities are associated with specific discharges using

$$Q = M + KS \qquad (2.21)$$

in which

M = mean
S = standard deviation
K = value taken from Table 2.4

The annual mean discharges for the Little Blue River (Appendix C), for example, have the parameters

Mean = M = 123.1 cfs
Standard deviation = S = 79.7 cfs

as computed in Table 2.5. If it is assumed that the annual mean discharges fit a normal distribution, the particular discharge that will be equalled or exceeded with a 10% probability can be computed: 10% exceedance probability, K = 1.282 (Table 2.4).

TABLE 2.5. Mean annual discharges, Little Blue River

Year	Q (cfs)	Q^2 (cfs)^2
1948	. . .	. . .
1949	168	28,244
1950	101	10,201
1951	176	30,976
1952	127	16,129
1953	40.2	1,616
1954	26.1	681
1955	59.5	3,540
1956	11.5	132
1957	22.8	520
1958	131	17,161
1959	45.6	2,079
1960	83.3	6,939
1961	279	77,841
1962	202	40,804
1963	43.7	1,910
1964	70.1	4,914
1965	162	26,244
1966	119	14,161
1967	185	34,225
1968	150	22,500
1969	257	66,049
1970	264	69,696
1971	107	11,449
Total	2830.8	488,011

$$\text{Mean} = \Sigma Q / n = 2830.8/23 = 123.1$$

$$\text{Standard deviation} = \left[\frac{\Sigma Q^2 - (1/n)(\Sigma Q)^2}{n-1} \right]^{1/2}$$

$$= \left[\frac{488,011 - (1/23)(2830.8)^2}{23-1} \right]^{1/2} = 79.7$$

$$Q_{10} = M + KS = 123.1 + 1.282(79.7)$$

or

$$Q_{10} = 225.2 \text{ cfs}$$

Several points can be computed and displayed graphically; arithmetic probability paper is suited to this purpose. The result will be a straight line as shown in Figure 2.9.

LOG PEARSON TYPE III DISTRIBUTION

Several curves have been proposed to fit the annual maximum streamflows. Federal agencies (4, 5) have agreed to standardize on the log Pearson type III curve; i.e., the logarithms of the discharges are assumed to fit a particular family of curves devised by Karl Pearson. The equation of the curve is not necessary; tables of values are used instead and are given in Table 2.6.

Steps in establishing the curve follow.

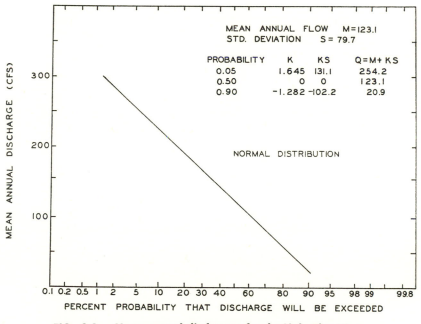

FIG. 2.9.   **Mean annual discharges for the Little Blue River.**

TABLE 2.6. Values of k in the Pearson Type III distribution

Skew coefficient (g)	Exceedance probability					
	0.99	0.90	0.50	0.10	0.02	0.01
3.0	—0.667	—0.660	—0.396	1.180	3.152	4.051
2.5	—0.799	—0.771	—0.360	1.250	3.048	3.845
2.0	—0.990	—0.895	—0.307	1.302	2.912	3.605
1.5	—1.256	—1.018	—0.240	1.333	2.743	3.330
1.2	—1.449	—1.086	—0.195	1.340	2.626	3.149
1.0	—1.588	—1.128	—0.164	1.340	2.542	3.022
0.9	—1.660	—1.147	—0.148	1.339	2.498	2.957
0.8	—1.733	—1.166	—0.132	1.336	2.453	2.891
0.7	—1.806	—1.183	—0.116	1.333	2.407	2.824
0.6	—1.880	—1.200	—0.099	1.328	2.359	2.755
0.5	—1.955	—1.216	—0.083	1.323	2.311	2.686
0.4	—2.029	—1.231	—0.066	1.317	2.261	2.615
0.3	—2.104	—1.245	—0.050	1.309	2.211	2.544
0.2	—2.178	—1.258	—0.033	1.301	2.159	2.472
0.1	—2.252	—1.270	—0.017	1.292	2.107	2.400
0.0	—2.326	—1.282	0.000	1.282	2.054	2.326
—0.1	—2.400	—1.292	0.017	1.270	2.000	2.252
—0.2	—2.472	—1.301	0.033	1.258	1.945	2.178
—0.3	—2.544	—1.309	0.050	1.245	1.890	2.104
—0.4	—2.615	—1.317	0.066	1.231	1.834	2.029
—0.5	—2.686	—1.323	0.083	1.216	1.777	1.955
—0.6	—2.755	—1.328	0.099	1.200	1.720	1.880
—0.7	—2.824	—1.333	0.116	1.183	1.663	1.806
—0.8	—2.891	—1.336	0.132	1.166	1.606	1.733
—0.9	—2.957	—1.339	0.148	1.147	1.549	1.660
—1.0	—3.022	—1.340	0.164	1.128	1.492	1.388
—1.2	—3.149	—1.340	0.195	1.086	1.379	1.449
—1.5	—3.330	—1.333	0.240	1.018	1.217	1.256
—2.0	—3.605	—1.302	0.307	0.895	0.980	0.990
—2.5	—3.845	—1.250	0.360	0.771	0.798	0.799
—3.0	—4.051	—1.180	0.396	0.660	0.666	0.667

1. Transform the discharges to their logarithms (base 10).

$$X = \log Q \tag{2.22}$$

2. Compute the mean of the logarithms (see Eq. 2.17).

$$M = (\Sigma X)/n$$

3. Compute the standard deviation of the logarithms (see Eq. 2.18).

$$S = \left[\frac{\sum (X^2) - \left(\sum X \right)^2 / n}{n - 1} \right]^{1/2}$$

4. Compute the skew of the logarithms (see Eq. 2.19).

$$g = \left[\frac{n^2 \left(\sum X^3 \right) - 3n \left(\sum X \right) \left(\sum X^2 \right) + 2 \left(\sum X \right)^3}{n(n-1)(n-2)S^3} \right]^{1/3}$$

5. Compute the curve from the relationship

$$\log Q = M + KS \qquad\qquad (2.23)$$

where K is selected from Table (2.6). K depends upon both the probability and the skew.

As an example, consider the annual maximums of the Little Blue River (Table 2.7); only the largest peak discharge for each year is used. The peak discharges were transformed by taking the common logarithms. The mean, standard deviation, and skew of the transformed variables were found to be

$$M = 3.541, \qquad S = 0.309, \qquad g = -1.91$$

The peak discharge, with a 10% probability of being exceeded, can be determined. From Table 2.6, using exceedance probability $= 0.1$, K is found by interpolation to be 1.088. Thus

$$X_{10} = \log Q_{10} = 3.541 + 1.088\,(0.309) = 3.877$$

and

$$Q_{10} = 10^{3.877} = 7533 \text{ cfs}$$

several points may be computed and the results displayed graphically as shown in Figure 2.10. Logarithmic probability paper is suited to this application.

The log Pearson type III distribution reduces to another distribution, the log-normal distribution, when the skew is zero. The log-normal is obtained by transforming the random variable by taking its logarithm and using the normal distribution with the transformed variables. The log-normal and, obviously, the log Pearson type III with zero skew plot as straight lines on logarithmic probability paper.

An excellent discussion of the log-normal distribution is given by Yuan (6), and applications to hydrologic data are given by Chow (7). A discussion of the Pearson type III, without the logarithmic transformation, and its applications to hydrologic data are given by Foster (8).

TABLE 2.7. Annual peak discharges, Little Blue River

Year	Peak discharge Q (cfs)	Log $Q = X$	X^2	X^3
1948	6,000	3.778	14.274	53.931
1949	2,800	3.447	11.883	40.962
1950	5,580	3.747	14.037	52.592
1951	6,400	3.806	14.487	55.140
1952	3,690	3.567	12.724	45.386
1953	2,140	3.330	11.092	36.940
1954	2,820	3.450	11.904	41.073
1955	4,000	3.602	12.975	46.736
1956	408	2.611	6.816	17.793
1957	1,680	3.225	10.403	33.552
1958	4,350	3.638	13.239	48.169
1959	1,290	3.111	9.676	30.097
1960	2,600	3.415	11.662	39.826
1961	9,460	3.976	15.808	62.850
1962	4,640	3.667	13.443	49.290
1963	1,900	3.279	10.750	35.247
1964	2,240	3.350	11.224	37.604
1965	5,200	3.716	13.809	51.313
1966	8,000	3.903	15.234	59.460
1967	5,410	3.733	13.937	52.029
1968	2,270	3.356	11.263	37.799
1969	5,600	3.748	14.049	52.658
1970	9,450	3.975	15.804	62.828
1971	3,500	3.544	12.560	44.515
	$\Sigma =$	84.974	303.053	1,087.790

Mean (log Q) $= \Sigma X/n = 84.974/24 = 3.541$

Standard deviation (log Q) $= \left[\dfrac{\Sigma (X^2) - 1/n \ (\Sigma X)^2}{n-1} \right]^{1/2}$

$$= \left[\dfrac{303.053 - 1/24 \ (84.974)^2}{23} \right]^{1/2} = 0.309$$

Skew (log Q) $= \dfrac{n^2 \ (\Sigma X^3) - 3n \ (\Sigma X) \ (\Sigma X^2) + 2\Sigma X^3}{n \ (n-1) \ (n-2) \ S^3}$

$$= \dfrac{24^2 \ (1087.79) - 3 \ (24) \ (84.974) \ (303.053) + 2 \ (84.974)^3}{24 \ (23) \ (22) \ (0.309)^3}$$

$$= -1.191$$

GRAPHIC METHODS

The preceding material has been devoted to analytically fitting a curve to a set of data. Graphic curve fitting is also widely used; several methods are in use, but they all start with the same set of information. For annual maximum flows the discharges are ordered in terms of magnitude. The largest discharge is given the order number 1 and the smallest is given order number N, where N is the number of years of rec-

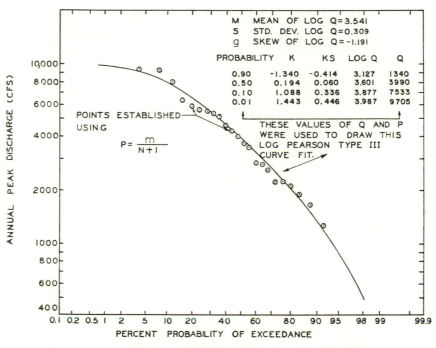

FIG. 2.10.   Peak annual discharges for the Little Blue River.

ord. Plotting positions p for the probability axis of the graph are assigned to each flow, the plotting position being determined only by the
order number m and the number of years of record. For example, the
largest flood in 50 years of record might be assigned the plotting position
of 1 in 50 if the formula

$$p = m/N \qquad\qquad (2.24)$$

happened to be used. Equation (2.24) represents one of several widely
used alternative formulas for assigning plotting positions.

After assigning a plotting position to each of the recorded discharges,
the results are graphed and a line of best fit drawn through the points.
The line is assumed to give the probabilities of all flows. The proper
formula to use in assigning plotting positions is not completely agreed
upon. The reasons for this difficulty can be illustrated by considering
2000 years of record during which hydrologic conditions do not change.
This can be broken into 100 segments of 20 years. The 20-year records
are typical of the period of record usually available for analysis. In
each of the 20-year records the peak discharge for each year is recorded.

The peak discharges for each 20-year record are ordered. Only one, the largest, flow for each year is used; hence this is termed an annual series. One hundred separate annual series are taken from the total 2000 years of record.

The largest flow in each segment will have order number $m = 1$, and the years of record N for each segment will be 20. The probability assigned to the largest flow of each segment will be the same, using any formula involving only m and N. Beard (9) indicates, "from knowledge of probability, it is expected that (1) one record will contain a flood that will be exceeded on an average of once in 2000 years, (2) 18 will contain 100-year floods (some records may contain more than one or the number would be 20), (3) 64 will contain floods larger than the 20-year flow, and (4) 12 will contain no floods larger than the 10-year flood." The plotting formula, however, will assign the same plotting position to the largest of the flows in each record. It is then appropriate to search for a plotting formula that will fit the data with a desired criterion of correctness.

One criterion would be to attempt to be correct on the average. The formula for this case is

$$p = m/(N + 1) \qquad (2.25)$$

Another criterion is to try to hit the median; i.e., half of the time the probability would be too low and half the time, too high. The probability for order number 1 is given by

$$p = 1 - 0.5^{1/n} \qquad (2.26)$$

An approximation applicable to all order numbers is

$$p = (m - 0.3)/(N + 0.4) \qquad (2.27)$$

We may attempt to be correct most of the time, i.e., use the modal value. The probability for the flow with order number 1 is given by

$$p = 1 - (1/e)^{1/n} \qquad (2.28)$$

Other plotting formulas have been proposed and are used; Eq. (2.25) is an example. The attempt to be correct on the average, Eq. (2.25), is probably the most widely used method. For the Little Blue River the largest peak flow is 9460 cfs. Its order number is 1 and the number of years of record is 24; therefore

$$p_{9460} = 1/(24 + 1) = 0.04 = 4\%$$

That is, a plotting position of 4% has been assigned to the likelihood that 9460 cfs will be equalled or exceeded.

As indicated, the discharges are each assigned a plotting position, the results graphed (logarithmic probability paper is sometimes used for flood discharges), and a curve drawn through the points. The plotted points for tht Little Blue River are compared with the log Pearson type III distribution in Figure 2.10.

One bit of information that may be included in the graphic approach is a historical flood that occurred before continuous records were kept. For example, the record may give information as shown in Table 2.8. Thus it is known that Q_2 is the largest flood in 65 years and Q_1 is the second largest; therefore, the associated probabilities are

$$P_{Q_2} = 1/66, \qquad P_{Q_1} = 2/66$$

The largest flood Q_3 in the continuous record may or may not be the third largest flood in the 65 years. There is no way of knowing. It is known, however, that it is the largest in 35 years; therefore, the appropriate probability is

$$P_{Q_3} = 1/36$$

PARTIAL DURATION SERIES

An alternate graphic approach, used on occasion, considers all discharges above some base discharge. The use of this approach arises because one year may have several large flows, whereas other years may have no flow even close to the magnitude of the large values. The annual series described earlier would consider only one of the large flows, neglecting the others entirely.

TABLE 2.8. Historical floods

Year	Discharge	Remarks
1900	Q_1	second largest known flood
1910	Q_2	largest known flood
1930–65		continuous record

TABLE 2.9. Corresponding values for annual and partial duration series

Annual series		Partial duration series	
Probability of exceedance (%)	Return period (yrs)	Interval between exceedances (yrs)	Number/ 100 yrs
86.4	1.16	0.5	200
63.4	1.58	1.0	100
50.0	2.0	1.45	69
39.4	2.54	2.0	50
18.2	5.52	5.0	20
9.5	10.5	10.0	10
1.98	50.5	50.0	2
0.99	100.5	100.0	1

Source: Langbein (10).

A series of discharges using all flows above some base is termed a partial duration series. The plotting formula usually used for this series is

$$p_e = m/N \qquad\qquad (2.29)$$

where m is the order number and N the number of years of record. The p_e can be interpreted as the average number of times per year that a given flow will be equalled or exceeded. The reciprocal of p_e can be interpreted as the average number of years between flows that equal or exceed the given discharge.

The partial duration series is probably the most useful in making an economic analysis, but frequently much greater effort is required to obtain the necessary data. As a result, the annual maxima are used and the results converted to a partial duration series, using Table 2.9 (6). Note too that the partial duration series approaches the annual series for the less frequent events.

DURATION CURVES

A modification of the partial duration series, called simply the duration curve, is regularly used in problems involving low flows. The duration curve is constructed by ordering the annual, monthly, weekly, or daily mean flows according to magnitude. The purpose of the study governs the choice of time interval. An order of magnitude is assigned to each event, as in previously discussed methods, with the largest event assigned the order 1. The assigned order number divided by the number of records ordered (the order number of the smallest flow in the group)

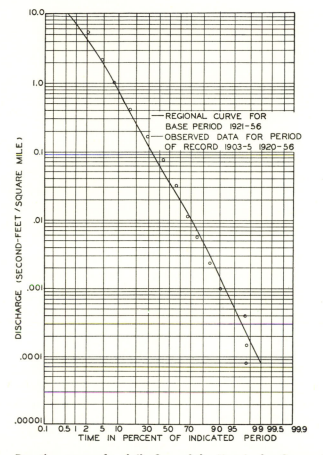

FIG. 2.11. Duration curves for daily flow of the Marais des Cygnes River (14).

and multiplied by 100 yields the percent of time intervals that a particular mean flow has been equalled or exceeded. An example of a duration curve is shown in Figure 2.11 (14).

The duration curve is frequently used to establish the water supply potential for municipal and industrial use, hydroelectric power, irrigation, and navigation where low flows are of importance. For example, if a certain municipal use at Ottawa, Kans., required a firm flow of 0.1 cfs/square mile, Figure 2.11 indicates that flow rate will be equalled or exceeded only 32% of the time. Thus the project planner must consider the possibility that flow requirements will not be met 68% of the time. Supplementary supplies would be required if the need is to be fulfilled.

CONCLUSION

This brief chapter devoted to probability and statistics does not provide skills for extensive analysis in the area. Further reading is necessary for the engineer who will be extensively involved in decision making with the use of probability and statistics. Benjamin and Cornell (11) have written a comprehensive treatment of applications in civil engineering, and their book is recommended for further study by the serious student.

More extensive details of methods of flood frequency analysis have been presented by Beard (9), Dalrymple (12), and Benson (13). An interesting analysis of a theoretical 1000-year record is given by M. A. Benson in (14).

PROBLEMS

2.1. What is the probability of a 2000-year flood occurring within any 30-year period?

2.2. The following parameters were determined for a series of annual maximum stream flows:

$$\text{Mean of log } Q = 3.00$$
$$\text{Standard deviation of log } Q = 0.392$$
$$\text{Skew of log } Q = 0.0$$

a. What is the discharge that has a 2% chance of occurring in any year?
b. If the design lifetime of the project is 25 years, what is the approximate probability that the discharge of part (a) will be exceeded in its lifetime?

2.3. If a project has a design lifetime of 100 years and a discharge with a 100-year return period is taken as the design discharge, what is the probability of failure in the lifetime of the project?

2.4. Fit a normal distribution to the mean annual flows of a river in your vicinity, or use one of the records given in the Appendices.

2.5. Fit a log Pearson type III distribution to the annual peak discharges of a river in your vicinity, or use one of the records given in the Appendices.

2.6. Assign probabilities to the annual peak discharges of the river chosen in Problem 2.5 and compare with the results of the log Pearson type III fit.

2.7. Imagine that during the next 2000 years someone records a continuous record of the annual flood flows. At the end of that period he constructs a log Pearson type III flood frequency diagram and compares it to one constructed by you on the basis of the 20-year record available to you.

  a. In what ways would you expect your diagram and his to be different?

  b. Which diagram would most accurately represent the true frequencies of occurrence?

  c. Explain why you in your study would or would not have confidence in the magnitude of flood predicted by each of the following exceedance probabilities: 98%, 50%, 0.1%.

2.8. The following figures represent annual peak flood magnitudes. Order the events. Construct a log Pearson type III plot of the events. Plot the given flood flows on your log Pearson type III plot according to apparent probability.

Year	Flood flow (cfs)	Year	Flood flow (cfs)
1945	1100	1957	300
1946	800	1958	2600
1947	400	1959	100
1948	1200	1960	900
1949	3300	1961	2100
1950	1100	1962	2000
1951	700	1963	700
1952	900	1964	400
1953	600	1965	3010
1954	200	1966	1100
1955	1300	1967	900
1956	1800	1968	800

2.9. Consider only the flows up through 1956 in Problem 2.8. What is the apparent probability of a peak flood of 2900 cfs? Consider all the records given in Problem 2.8. Now what is the apparent probability of a flow of 2900 cfs?

2.10. Construct a log Pearson type III plot for the record up through 1956 in Problem 2.8. What would the indicated value of a 200-year flood be? How does this compare with the 200-year flood given by the results of Problem 2.9?

REFERENCES

1. U.S. Weather Bureau. Hourly precipitation data, 1951–. Washington, D.C. (Climatological data, 1948–51; hydrologic bulletin, 1940–48.)
2. U.S. Geological Survey. Water supply papers. Washington, D.C.
3. U.S. Weather Bureau. Daily river stages. Washington, D.C.

4. Benson, M. A. Uniform flood-frequency estimating methods for federal agencies. *Water Resour. Res.,* Vol. 4, No. 5, Oct. 1968.
5. Water Resources Council. A uniform technique for determining flood flow frequencies. Bull. 15, Washington, D.C., Dec. 1967.
6. Yuan, P. T. Logarithmic frequency distribution. *Ann. Math. Stat.,* Feb. 1933.
7. Chow, Ven Te. The log-probability law and its engineering applications. Proc. ASCE, Vol. 80, No. 536, Nov. 1954.
8. Foster, H. A. Theoretical frequency curves and their application to engineering problems. Trans. ASCE, Vol. 87, 1924.
9. Beard, L. R. *Statistical Methods in Hydrology.* U.S. Army, Corps of Engineers, 1962.
10. Langbein, W. B. Annual floods and the partial duration flood series. *In* Trans. Am. Geophys. Union, Vol. 30, Dec. 1949.
11. Benjamin, J. R.; and Cornell, C. A. *Probability, Statistics and Decision for Civil Engineers.* McGraw-Hill, 1970.
12. Dalrymple, T. Flood-frequency analysis. U.S. Geological Survey Water Supply Paper 1543-A, 1960.
13. Benson, M. A. Evolution of methods for evaluating the occurrence of floods. U.S. Geological Survey Water Supply Paper 1580-A, 1962.
14. Furness, L. W. Kansas streamflow characteristics. 1. Flow duration. Kansas Water Resources Board, 1959.

CHAPTER THREE

APPLICATION OF PROBABILITY AND STATISTICS TO SELECTION OF A DESIGN DISCHARGE

DESIGN OF ANY FACILITY whose operation involves streamflows, whether they are flood flows or daily mean flows, involves some risk. The likelihood that a specific design discharge will be exceeded one or more times in the life of a structure was determined in the preceding chapter. If a structure will be damaged by a flow larger than that for which it is designed or if a project will suffer if the mean annual flow is smaller than the design value, some uncertainty must be accepted. How is the acceptable level of risk established? That question was implied in Chapter 2. In this chapter the manner in which the cost of risk may be considered will be explored. The principles will be established through examples: the first is an evaluation of alternate levee designs, the second is selection of the design discharge for a culvert.

EVALUATION OF ALTERNATE LEVEE DESIGNS

Assume that a project is being planned that will involve flood protection through use of levees. Three alternative levee designs are proposed and their costs are estimated. Each design provides a

TABLE 3.1. Costs of alternate levee systems

Level of protection (cfs)	Annual cost of project (50-yr life)
31,000	$ 31,700
48,000	36,800
64,500	45,300

measure of protection against a different flood level. The damages prevented must exceed the cost of the levees if the project is to be economically feasible. The estimated cost of each alternative and the level of protection are listed in Table 3.1.

The study of topography, property value, potential damage, and projected land use and the analysis of river hydraulics yields river discharge as a function of river stage and expected damages occurring at that stage without flood protection. Table 3.2 summarizes these figures.

Probability of occurrence of any particular flood stage must be established. This is accomplished by constructing (using the methods of Chapter 2) the partial duration series shown in Figure 3.1. Consider the levee system designed to protect against any flood less than 31,000 cfs. Assume for computational purposes that the period of analysis is 100 years. The average number of times a flood of any given discharge will be exceeded in 100 years can be read directly from the partial duration series (Fig. 3.1). Thus a discharge of 1900 cfs will be exceeded 61 times on the average, whereas a discharge of 2900 cfs will only be exceeded 39 times. These values are tabulated in columns 2 and 3 of Table 3.3. Thus we can expect there will be $61 - 39 = 22$ floods in 100 years whose discharges are between 1900 cfs and 2900 cfs.

The damage resulting from a flood of 1900 cfs is zero, but that resulting from a flood of 2900 cfs is $6000. Some of the 22 floods that are expected to occur within this range will cause almost no damage. We might consider that the average damage is $(0 + \$6000)/2 = \$3000$;

TABLE 3.2. Stage-discharge-damage relationship

Stage (ft)	Discharge (cfs)	Expected damage if discharge occurs
18	1,900	$ 0
19	2,900	6,000
20	6,100	39,000
21	15,500	213,000
22	33,000	767,000
23	70,000	1,005,000
24	144,000	1,065,000

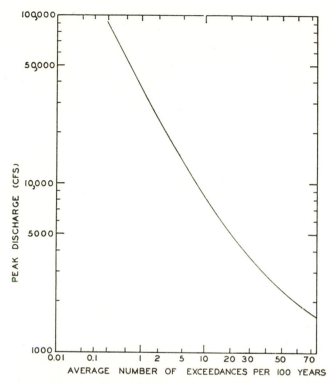

FIG. 3.1. Partial duration series for economic study of a levee system.

TABLE 3.3. Computation of average damages

Stage (ft)	Discharge (cfs)	Average exceedances in 100 years	Average exceedances within range	Average damage for range	Probable 100-yr damage cost for each range
(1)	(2)	(3)	(4)	(5)	(6)
18	1,900	61			
			22	$ 3,000	$ 66,000
19	2,900	39			
			23	22,500	517,500
20	6,100	16			
			11.9	126,000	1,500,000
21	15,500	4.1			
			2.8	466,000	1,300,000
21.9	31,000	1.3			
			0.1	743,000	74,300
22	33,000	1.2			
			0.58	826,500	480,000
22.5	48,000	0.62			
			0.25	933,500	233,000
22.9	64,500	0.37			
			0.05	993,000	49,600
23	70,000	0.32			

thus the expected damage for 100 years due to floods between the discharges 1900 cfs and 2900 cfs is 22($3000) = $66,000. Average damages are listed in column 5 of Table 3.3, and expected 100-year damages are shown in column 6. In the 100-year computational period 39 floods will exceed a discharge of 200 cfs. Expected damages from these floods are computed in the same manner and constitute the lower portion of Table 3.3.

The total of damages prevented by a levee protecting up to the 31,000 cfs flood level is the sum of the damages from smaller floods. Thus the damages prevented in 100 years would be $66,000 + $517,000 + $1,500,000 + $1,300,000 = $3,383,500. Usually we would convert this to average damages prevented each year. Since the floods would occur at random throughout the 100 years, the total damages prevented are simply divided by 100 to get average annual values. Average annual damages prevented by the other alternative levee designs are determined in the same manner.

The three alternative levee systems can now be compared economically. If the levee system were a portion of a competitive market, the optimal economic level of development would occur when the incremental benefits (column 6) just equal the incremental costs and the benefits exceed the cost. This is not necessarily true of a monopolistic situation, which is the case for the levee. The approach considered appropriate by the U.S. federal government is to act as though a competitive market exists and to develop to the point where incremental benefits equal the incremental costs, recognizing that due to the monopolistic situation the maximum difference between benefits and cost may occur at a lower level of development.

Table 3.4 summarizes the computations for the three possibilities under consideration. The step from protection at the 31,000 cfs level to the 48,000 cfs level provides $5543 in benefits but costs only $5100 on an annual basis. To make the next step, however, $8500 is spent to gain only $2330 in benefits. Obviously, we should stop at the 48,000 cfs level.

TABLE 3.4. Benefit-cost analysis

Level of protection (cfs)	Annual benefits	Incremental benefits	Annual project costs	Incremental costs	Benefits minus costs
31,000	$33,835		$31,700		$2,135
		$5,543		$5,100	
48,000	39,378		36,800		2,578
		2,330		8,500	
64,500	41,708		45,300		—3,592

In this particular case the difference between the benefits and costs is also a maximum.

It is of interest to note that annual dollar value of flood damage is increasing in the United States even though extensive flood control projects have been completed in the last 40 years. This is due to the increased level of development that seems to occur when some level of flood protection is provided. Recognition of this phenomenon should be utilized in preparing the expected damage values of Table 3.2.

SELECTION OF CULVERT SIZE

As a second example of choosing a design flood, an analysis aimed at selection of an appropriate roadway culvert diameter is given. The data used in this section are adapted from a report by Pritchett (1). In this analysis the benefits will not be compared with the costs; that would be done for the total road. In this case the goal is to minimize the culvert cost, which is made up of the initial cost, regular maintenance cost, and the cost of any damages resulting from insufficiency of the structure.

Consider a circular, 68-foot reinforced concrete pipe culvert to be constructed on a stream for which the partial duration series for peak discharges is as given in Table 3.5. The construction costs of the concrete pipe and of the headwalls computed on an annual investment basis are given in Figure 3.2 (1). These annual costs are computed assuming a 7% annual rate of interest and a 30-year life for the structure. The cost of annual maintenance is estimated to be $6.00 and is assumed to be independent of the size of the culvert.

The pipe diameters listed in Table 3.5 are the sizes required to pass the specified flow rate without overtopping the roadway or otherwise exceeding design conditions. The necessary pipe sizes were computed by using hydraulic principles and knowledge of the particular site.

TABLE 3.5. Required culvert sizes and probability of flow exceedance

Discharge (cfs)	Culvert diameter required to pass discharge (in.)	Probable number of exceedances per year
21	21	0.33
24	24	0.20
28	24	0.10
32	27	0.05
35	30	0.025
40	30	0.01

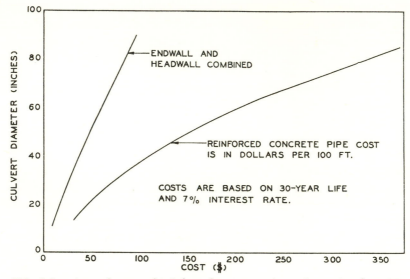

FIG. 3.2. Annual costs of reinforced concrete pipe culverts in place (1).

Because only commercially available pipe sizes were considered, the same required pipe diameter appears for more than one discharge.

For this particular culvert the damages are assumed to be given by

$$\text{Damage costs} = \$50 + \$6(Q - Q_D) \tag{3.1}$$

for discharges Q greater than the design discharge Q_D, but less than the discharge causing destruction of the structure.

Destruction is assumed to occur from storms with a return period equal to or greater than 100 years, and the cost of repairs plus other costs is assumed to vary linearly from the 100-year cost to a maximum of $2000 plus replacement cost.

The detailed computations for a 24-inch culvert are illustrated in Table 3.6. Initial cost (on an annual investment basis) is obtained from Figure 3.2. Annual maintenance is noted, and a table is constructed for computing the probable annual damage costs. Probable annual damage is obtained as the product of columns 3 and 4 of Table 3.6. The total probable annual cost for this culvert is then

$$\text{Total probable annual cost} = \$58.20 + \$6.00 + \$24.62 = \$88.82$$

A similar analysis should be carried out for other culvert sizes and the results compared. Results are given in Figure 3.3 (1). A minimum annual cost will be obtained for a 24-inch diameter culvert.

TABLE 3.6. Analysis of a 24-inch diameter, reinforced concrete culvert, design discharge = 28 cfs

	Initial cost:			
	Pipe	$57(68/100)	= $38.80/year	
	Headwalls and endwalls		= 19.40/year	
	Total initial cost		= $58.20/year	
	Maintenance		= $ 6.00/year	

Discharge (cfs)	Exceedances per year	Number of exceedances per year within range	Average damages for range	Probable annual damages
(1)	(2)	(3)	(4)	(5)
28	0.10			
		0.05	$ 62	$ 3.10
32	0.05			
		0.025	83	2.08
35	0.025			
		0.015	107	1.61
40	0.01			
		0.01	1,783*	17.83
Greater	0.00			
Total probable annual damage				$24.62

* Average of 100-year flood damage and maximum damage plus present worth replacement cost of culvert.

$$\text{Average damage} = \tfrac{1}{2}\left\{\,[50 + 6(40-28)] + 2{,}000\,\right\}$$
$$+ 58.20(12.409) = 1{,}783$$

in which 12.409 is the series present worth factor for 7% interest over 30 years.

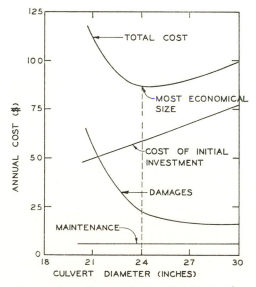

FIG. 3.3. Comparison of annual costs of various culvert sizes (1).

TABLE 3.7. **Average design frequencies for highway drainage structures**

Type of highway	Type of structure	Average design frequency (yrs)
Secondary	small	24
	large	35
Primary	small	37
	large	47
Interstate	small	47
	large	50+

Source: Wycoff and Harbaugh (2).

Small highway drainage structures are regularly considered to be sufficiently minor in cost that the savings resulting from an extensive analysis does not justify the cost. A survey of state highway department practice was made by Wycoff and Harbaugh, the results of which are summarized in Table 3.7 (2).

The average design discharge for a small structure on a primary highway is that with a return period of 37 years according to Table 3.7. For the example in the preceding section this would call for a 27-inch culvert. The saving resulting from the economic anaylsis would be $4.00 per year. The fairly close agreement of the economic analysis with that obtained using average values probably is an example of success through trial and error. A vast number of culverts have been built, and this experience results in nonquantified standards in close agreement with quantified results. The foregoing is not meant to detract from the desirability of careful analysis, but is meant to indicate that the extent of the analysis should fit the problem at hand.

CONCLUSION

The preceding analyses indicate the combined use of statistical and economic data in selecting the level of development. Procedures of this type are very useful but must be carefully applied. Many costs and benefits are difficult to include in the analysis. It is difficult, for example, to assign a monetary value to aesthetics. The loss of human life is likewise difficult to assess monetarily; in fact it is sometimes given an unbounded value, leading to design for the worst conceivable conditions.

Analysis of the economic aspects of water resource systems leads to many interesting problems in both hydrology and economics. The problems are explored in more detail in references (3), (4), (5), (6), (7), and (8).

PROBLEMS

3.1. A recording stream gage installation (gaging station) can be constructed on the bank of a particular river at an equivalent annual cost of $2000. The lifetime of the structure and recorder is six years. Maintenance, chart paper, and batteries to power the recorder cost $100/year. Major floods below 5000 cfs will not damage the installation, but floods greater than 5000 cfs will cause total destruction of the gaging station. The 5000 cfs flood has an exceedance probability of 1%. What is the total probable annual cost of this gaging station?

3.2. Tabulated below are flood stages, damages resulting from a flood of that stage, exceedance probabilities, and cost of protection against flooding. Determine the appropriate level of protection to provide.

Flood stage with no protection (ft)	Damage, each occurrence*	Percentage chance of flood exceeding	Incremental annual cost of protection
above 24	$12,000,000	0	
24	8,000,000	0.1	
			$80,000
22	3,000,000	0.3	
			38,000
21	2,000,000	1.0	
			32,000
20	1,500,000	2.3	
			27,000
19	1,000,000	4.3	
			23,000
18	600,000	7.6	
			18,000
17	400,000	12.6	
			13,000
16	200,000	19.3	
			17,000
15	100,000	29.3	
			3,000
14	0	49.3	

* Occuring between this magnitude and the magnitude below it.

3.3. Show that the tabular computation process of Table 3.3 is the numerical integration of a graph of damage plotted against probability.

REFERENCES

1. Pritchett, H. D. Application of principles of engineering economy to the selection of highway culverts. Institute in Engineering Economic Systems, Rept. EEP-13, Stanford Univ., 1964.
2. Wycoff, R. L.; and Harbaugh, T. E. A survey of hydraulic design practices of state highway departments. Hydrol. Ser. Bull., Univ. Mo., Rolla, June 1970.
3. Grant, E. L.; and Ireson, W. G. *Principles of Engineering Economy.* Ronald Press, 1960.
4. Morgali, J.; and Oglesby, C. H. Procedures for determining the most economical design for bridges and roadways crossing flood plains. Highway Research Board Bull. 320, Jan. 1962.
5. James, L. D.; and Lee, R. R. *Economics of Water Resources Planning.* McGraw-Hill, 1971.
6. Eckstein, O. *Water Resource Development, The Economics of Project Evaluation.* Harvard Univ. Press, 1958.
7. Kurtilla, J. V.; and Eckstein, O. *Multiple Purpose River Development.* Johns Hopkins Press, 1958.
8. Maass, A., et al. *The Design of Water Resource Systems.* Harvard Univ. Press, 1962.

CHAPTER FOUR
PRECIPITATION

IN PREVIOUS CHAPTERS the concepts of probability and statistics were introduced and applied to analysis of streamflow data. The use of this analysis in decision making was demonstrated. If adequate streamflow data were always available and streamflow were the only matter of interest, there would be no need to delve deeper into the fundamental aspects of hydrology. This is not the case, however, and it is frequently necessary to analyze precipitation in order to estimate streamflow, groundwater accretion, and other hydrologic quantities.

In the discussion of the hydrologic cycle in Chapter 1 it was indicated that streamflow results directly or indirectly from precipitation. Although this concept now seems obvious, it was the subject of considerable controversy in 1674 when Pierre Perrault's book *De l'Origine des Fontaines* (The Origin of Springs) was published in Paris. Perrault showed by example and the use of his own annual rainfall measurements that rainfall was indeed sufficient to provide for the flow of rivers (1). It was popularly believed during Perrault's time that the water flowing from springs was manufactured in some mysterious way in the depths of the earth (2).

ELEMENTS OF METEOROLOGY

Meteorology, the science of the study of the atmosphere, considers the related phenomena of wind, precipitation, temperature, pressure, and humidity. As a branch of physics, meteorology is concerned with the atmosphere as a mixture of gases for which the pressure, temperature, and volume interrelations follow the laws of thermodynamics. Geographic considerations are also involved because latitude, altitude, topography, and the location of land and water areas affect the character and

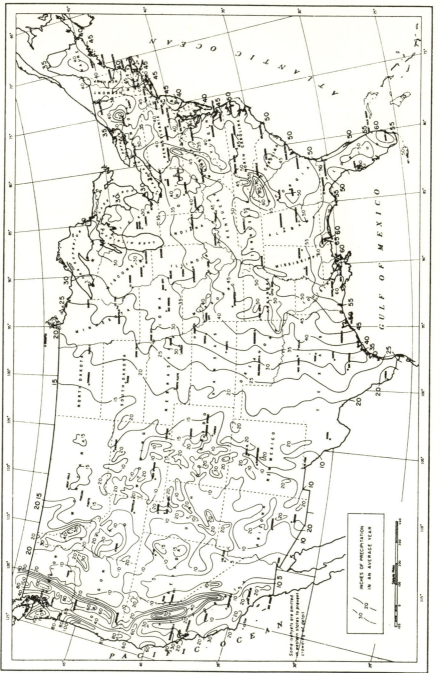

FIG. 4.1. Average annual precipitation of the United States (3).

distribution of meteorological conditions over the surface of the earth. All these conditions influence the amount of precipitation and how and when it occurs. Figure 4.1 shows the average annual precipitation for the United States (3).

CIRCULATION OF THE ATMOSPHERE

Energy involved in producing movement of the earth's atmosphere comes directly from the sun. Certain general patterns of this movement can be deduced from average conditions of temperature and pressure that exist within the atmosphere. In the vicinity of the equator the sun tends to warm the atmosphere to a greater degree than in nearby regions to the north and south. Since it is lighter than the cooler surrounding air, this warmed air rises. Conversely, a general condition of cold occurs at the earth's poles causing air to descend. This might be expected to result in surface winds blowing from the poles toward the equator, as illustrated by the single-loop model in Figure 4.2. However, the earth is rotating and is not uniformly heated around the equator. These factors greatly complicate the dynamic behavior of the atmosphere.

The air at the equator cools as it rises and moves toward the poles. This cooling takes place so rapidly that the air, rather than continuing aloft to the poles, descends at about 30°N latitude. The descending air creates a high-pressure area at the surface of the earth. As a result, air north (in the Northern Hemisphere) of the high-pressure region moves northerly, meeting the southerly flow from the polar region. Three separately circulating cells result, as shown on the right side of Figure 4.2.

An observer on the earth would see the winds blowing directly north and south if the earth were not rotating. The air moving southward from the north polar region initially has only a southerly velocity component. As a result it is left behind as the earth rotates. By established convention, a wind is named according to the direction from which it blows. Thus a north wind blows from the north and toward the south. To an observer standing on the earth the wind from the north pole appears to flow toward the southwest, creating a region of polar easterlies in the region above approximately 60°N latitude. In a similar fashion westerlies and easterlies result in the regions respectively north and south of the 30°N latitude as indicated in Figure 4.2. This apparent deviation of the velocity can be discussed in terms of the apparent Coriolis force as is normally done in textbooks on dynamics (4).

The discussion in the preceding paragraphs dealt only with the Northern Hemisphere. Similar conditions prevail in the Southern Hemisphere except that the Coriolis component of acceleration causes the

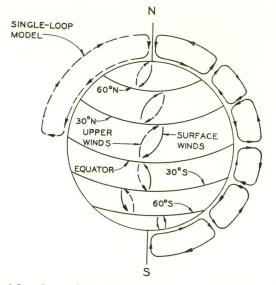

FIG. 4.2. General circulation of the earth's atmosphere.

surface winds to deviate to the left, resulting in the general directional pattern indicated in Figure 4.2. Winds completing the circulation of the atmosphere occur at high altitudes and are also shown.

Some average characteristics of pressure can be deduced from the circulation pattern of Figure 4.2. Most people are familiar with the low-pressure and high-pressure areas indicated on weather maps. Flow patterns associated with these areas are shown in Figure 4.3. Low pressure thus tends to dominate at the equator and at 60°N and S latitude, while high pressure tends to dominate conditions near 30°N and S latitude. Average pressures and wind directions are indicated in Figure 4.4. These are only average conditions, however, and considerable local variation exists due to seasonal changes and instability in the motion of the earth's atmosphere.

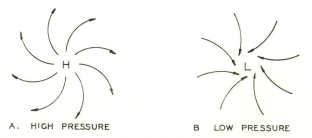

FIG. 4.3. Flow patterns associated with high-pressure and low-pressure centers near the earth's surface in the Northern Hemisphere.

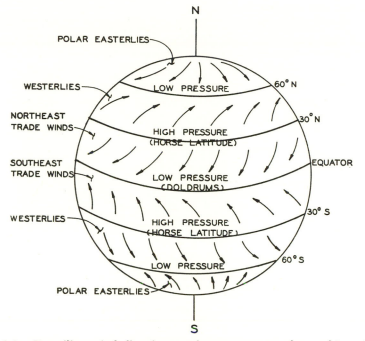

FIG. 4.4. Prevailing wind directions and pressures near the earth's surface.

Warm humid air is carried aloft near the equator. In rising, the air is cooled and produces condensation and considerable precipitation. As a result this region is one of considerable cloudiness, light and variable winds, frequent thunderstorms, squalls, calms, and high annual precipitation. Because of the frequent calms and cloudiness this region became known as the *doldrums* in sailing days.

In the region near 30°N and S latitude, where high pressure predominates, the great arid deserts of the earth are found. The air descending from above is, in general, quite dry. On the ocean these regions, marked by clear skies and low precipitation, became known as the *horse latitudes,* presumably because early sailing ships carrying horses to the New World were frequently becalmed there and were either forced to throw horses overboard or allow the crews to die of thirst. Similarly, the polar regions experience little snowfall because of the lack of moisture in the descending air.

The low-pressure regions at 60°N and S latitude are regions of rising warm air and experience moderate precipitation. Intense atmospheric movement occurs there, however (at least partially due to the large shearing forces in the atmosphere), and major storms frequently develop in these regions.

OROGRAPHIC AND SEASONAL EFFECTS

Land and ocean masses influence the circulation of the atmosphere to a significant degree. In winter, cold land masses tend to produce intense high-pressure areas, while low-pressure areas form over adjacent relatively warm oceans. As a result the prevailing wind in this season is dry and flows toward the ocean. In summer when the land is relatively warm as compared to the ocean, the high- and low-pressure patterns reverse and humid air flows in over the land, producing the *monsoon* rains. Northern India experiences extremely large amounts of this type of precipitation with almost all falling within the summer months. North Vietnam, South Vietnam, Laos, Thailand, and many other Asian countries also experience monsoons.

PRECIPITATION

Precipitation results from condensation of moisture in the atmosphere. Basically, three conditions must be met for condensation to occur and raindrops to form: (1) The conditions for saturation must be produced, usually through cooling; (2) water vapor must change phase to liquid and/or solid; and (3) small water droplets or ice crystals must grow to a size enabling them to fall. Clouds result from the occurrence of the first two steps, but often the precipitation never materializes. Artificial seeding of clouds with silver-iodide crystals has been frequently used in an attempt to produce the last step.

Cooling almost always takes place as a result of lifting. There are three lifting processes: convectional, orographic, and cyclonic.

Convectional lifting results from localized heating of the earth's surface. Warm air masses are lifted in cells, producing the thunderstorms so commonly experienced in afternoons of hot days. Precipitation from such storms is of short duration but may be quite intense.

Orographic lifting results when air is blown over mountains. Southern British Columbia is an excellent example of an area where precipitation occurs due to orographic effects. As much as 150 inches of rain falls annually on the seaward side of Vancouver Island where effective condensation occurs as air masses moving inland rise to pass over mountains. Because so much moisture content is lost on this initial lifting, the annual precipitation in the Kootenay Valley several hundred miles inland where another lifting process occurs is only some 50 inches (5).

Cyclonic lifting results from a warm light-air mass colliding with and riding up over a heavier cold-air mass. The low-pressure regions at 60°N and S latitude draw warm, somewhat humid air from the south and cool drier air from the north. These masses of air are quite well de-

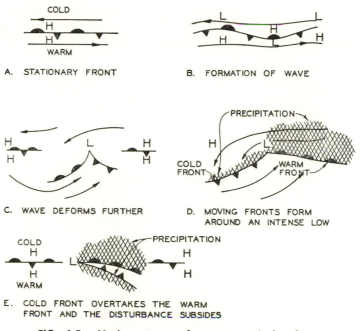

A. STATIONARY FRONT

B. FORMATION OF WAVE

C. WAVE DEFORMS FURTHER

D. MOVING FRONTS FORM AROUND AN INTENSE LOW

E. COLD FRONT OVERTAKES THE WARM FRONT AND THE DISTURBANCE SUBSIDES

FIG. 4.5.   Various stages of an extratropical cyclone.

fined, the boundary being delineated by a front, the Polar Front. If warm air is pushing out a cold air mass, the front is termed a warm front or if the opposite is true, a cold front.

The interface between these air masses tends to be unstable, and waves with associated highs and lows readily materialize. Figure 4.5 shows the history of the development of an extratropical cyclone at the polar front. The phases of the development are:

1. A stationary front exists with flow in opposite directions on each side of the front.
2. A disturbance produces a wave in the streamlines. The bending of the streamlines produces low- and high-pressure areas at the crest and trough of the wave respectively.
3. The new pressure distribution further deforms the front, which begins to wrap up, with the warm air riding over the heavier cold air.
4. As the warm air advances and rides over the cold air (warm front), precipitation forms over a wide area. Cold air advancing behind the moving warm air (cold front) produces a narrow band of precipitation along the cold front.

5. The fast-moving cold front overtakes the warm front, trapping warm air aloft. The original cause of the wave has been dissipated and the front returns to the initial condition (1). Further waves may develop.

This type of storm, referred to as an extratropical storm, may vary in size from one covering only a few hundred square miles to an intense low-pressure system covering tens of thousands of square miles. Such storms formed over the Atlantic and Pacific oceans are called *hurricanes* and *typhoons* respectively.

ANALYSIS OF PRECIPITATION AT A POINT

Precipitation is measured officially by the U.S. Weather Bureau by means of a highly standardized rain gage. A network of more than 14,000 of these gages is maintained largely by volunteer help. As a result, a great deal of data on precipitation is available for analysis (6). The measurements are recorded as depth of water falling within a given time. About 20% of the U.S. Weather Bureau rain gages record precipitation continuously, the remainder are read only once daily. Each rain gage catches only the precipitation falling within an 8-inch diameter circle; as a result, measurements by a single gage effectively represent precipitation only at a single point.

When point data are considered, two difficulties regularly arise. The catch for one rain gage may be missing due to failure to record it, temporary loss of the gage, or some other reason. At any rate, it is frequently desirable to estimate this missing value. Another problem results from moving a rain gage from one location to another, such as moving from an old airport to a new airport. The records must be modified if the catch at the new location is to be used in conjunction with the records of the old location. Both of these problems are met by comparison with nearby rain gages.

ESTIMATING A MISSING VALUE

To fill in a missing value representative of a long period of time, e.g., the total rainfall for a year, an average of ratios may be used.

$$P_x/\overline{P}_x = (1/N) \sum_{i=1}^{N} (P_i/\overline{P}_i)$$

where

P_x = unknown value
$\overline{P}_x$ = average annual catch at station with missing value

N = number of surrounding stations used to estimate missing value

P_i = catch for the same time period as the missing value but at station i

$\overline{P_i}$ = average annual catch at station i

For short time periods, if a value is missing for a particular storm, it is better to construct an isohyetal map as described in the next section and to interpolate a value from this map.

ESTABLISHING CONSISTENCY OF A RECORD

To establish the consistency or lack of consistency of a record, a double-mass analysis is used. The cumulative catch (Table 4.1) of the gage in question is plotted against the combined cumulative catch of several surrounding rain gages, as indicated in Figure 4.6 (7). If the record is consistent, the graph should plot as a straight line as shown for station A. If it is not, the graph will exhibit a change in slope as for station E.

To synthesize a continuous record, the observed data are adjusted by multiplying them by the ratio of the slopes of the two line segments:

$$P_a = (b_a/b_o)P_o$$

in which

P_a = adjusted precipitation
P_o = observed precipitation
b_a = slope of line to which records are adjusted
b_o = slope of line at observed P_o

The values for station E, 1926–30, are adjusted to be continuous with 1931–42 as shown in Table 4.2. Double-mass analysis is not suitable for adjusting daily or storm precipitation.

INTENSITY-DURATION ANALYSIS

Having established a continuous, consistent record, we can turn to the variability within the record. Figure 2.1 illustrates the way in which annual rainfall (total depth falling during the year) can vary during a long period. The variation from year to year effectively masks any long-term variation. Annual rainfall is described by its long-term mean, and the standard deviation is a measure of its variability.

Precipitation also varies with time within each particular storm, and the duration (total time during which rain falls) varies from storm to storm; therefore, analysis of precipitation at a point must involve both the amount (depth) of rain that falls and the elapsed time (duration) during which that amount fell. This is called intensity-duration analysis

TABLE 4.1. Double-mass analysis

Year	Annual precipitation for stations indicated (in.)						Cumulative annual precipitation for stations indicated (in.)					
	A	B	C	D	E	Mean	A	B	C	D	E	Mean
1926	39.75	45.70	30.69	37.36	32.85	37.27	39.75	45.70	30.69	37.36	32.85	37.27
1927	39.57	38.52	40.99	30.87	28.08	35.61	79.32	84.22	71.68	68.23	60.93	72.88
1928	42.01	48.26	40.44	42.00	33.51	41.24	121.33	132.48	112.12	110.23	94.44	114.12
1929	41.39	34.64	32.49	39.92	29.58	35.60	162.72	167.12	144.61	150.15	124.02	149.72
1930	31.55	45.13	36.72	36.32	23.76	34.70	194.27	212.25	181.33	186.47	147.78	184.42
1931	55.54	53.28	62.35	36.61	58.39	53.23	249.81	265.53	243.68	223.08	206.17	237.65
1932	48.11	40.08	47.85	38.61	46.24	44.18	297.92	305.61	291.53	261.69	252.41	281.83
1933	39.85	29.57	32.74	26.89	30.34	31.88	337.77	335.18	324.27	288.58	282.75	313.71
1934	45.40	41.68	36.13	32.44	46.78	40.49	383.17	376.86	360.40	321.02	329.53	354.20
1935	44.89	48.13	30.73	41.56	38.06	40.67	428.06	424.99	391.13	362.58	367.59	394.87
1936	32.64	39.48	35.40	31.32	42.82	36.33	460.70	464.47	426.53	393.90	410.41	431.20
1937	45.87	44.11	39.16	44.14	37.93	42.24	506.57	508.58	465.69	438.04	448.34	473.44
1938	46.05	38.94	43.27	50.62	50.67	45.91	552.62	547.52	508.96	488.66	499.01	519.35
1939	49.76	41.58	49.85	41.09	46.85	45.83	602.38	589.10	558.81	529.75	545.86	565.18
1940	47.26	49.66	47.86	39.01	50.52	46.86	649.64	638.76	606.67	568.76	596.38	612.04
1941	37.07	31.92	32.15	34.45	34.38	33.99	686.71	670.68	638.82	603.21	630.76	646.03
1942	45.89	38.16	52.39	47.32	47.60	46.27	732.60	708.84	691.21	650.53	678.36	692.30

Source: U.S. Geological Survey (7).

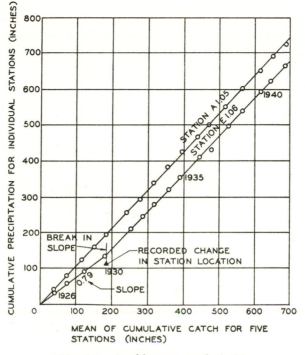

FIG. 4.6. Double-mass analysis (7).

and proceeds in the following manner. The rainfall record from a recording rain gage is listed in Table 4.3 (8). A particular duration is selected and the maximum rainfall for this time is determined. The maxima for all storms are listed in order of decreasing magnitude. Table 4.4 is an example of an analysis of a 10-minute duration rainfall for Chicago, Ill.; column 1 is the order number m, column 2 is the rainfall in the most intense 10 minutes y, and column 3 is the return period assigned

TABLE 4.2. Adjusted precipitation data for station E (1926–30)

Year	Original value (in.)	Adjusted value (in.)
1926	32.85	44.08
1927	28.08	37.68
1928	33.51	44.96
1929	29.58	39.69
1930	23.76	31.88

Source: U.S. Geological Survey (7).

TABLE 4.3. Excessive precipitation data (precipitation values are in inches)

Date	Year	Duration (min)													
		5	10	15	20	25	30	35	40	45	50	60	80	100	120
July 14	1913	0.16	0.29	0.40	0.50	0.59	0.67	0.74	0.79						
Aug. 7		0.31	0.37												
Aug. 7–8		0.30	0.44	0.56											
Aug. 18		0.28	0.49	0.63	0.67										
Apr. 27	1914	0.27													
May 27		0.18	0.33	0.41	0.49										
June 4		0.21	0.35	0.40											
July 16		0.33	0.66	0.79	0.97	1.21	1.48	1.61							
Aug. 9		0.35	0.62	0.83	0.91										
Aug. 13		0.19	0.36	0.50	0.60	0.68									
Sept. 1		0.14	0.27	0.38	0.40										
May 15	1915	0.17	0.25	0.32	0.40	0.48									
June 12		0.18	0.31	0.46	0.56	0.76	0.82	0.89	0.92	0.98					

(Data from July 7, 1915, through June 27, 1946, was listed and analyzed but is not shown here.)

Date	Year	5	10	15	20	25	30	35	40	45	50	60	80	100	120
June 30	1946	0.40	0.40	0.40	0.40	0.40	0.40	0.40	0.40	0.41	0.41	0.43	0.50	0.57	0.73
July 9		0.37	0.63	0.89	0.97	1.12	1.26	1.43	1.60	1.76	1.95	2.15	2.35	2.40	2.43
Aug. 9		0.25	0.50	0.63	0.70	0.75	0.78	0.79	0.80	0.80	0.80	0.80	0.85	0.95	1.13
Apr. 4–5	1947	0.33	0.58	0.73	0.86	0.98	1.06	1.10	1.15	1.19	1.26	1.48	1.53	1.75	1.91
July 6		0.38	0.60	0.76	0.86	1.10	1.29	1.43	1.55	1.66	1.71	1.92	2.28	2.32	2.37
July 13		0.31	0.44	0.57	0.62	0.64	0.66	0.68	0.70	0.72	0.73	0.75	0.81	0.84	0.84
Aug. 29		0.36	0.60	0.72	0.77	0.81	0.83	0.85	0.87	0.89	0.90	0.91			
Sept. 11		0.25	0.50	0.72	0.77	0.77	0.78								
Sept. 21		0.13	0.23	0.33	0.42	0.50	0.57	0.62	0.67	0.72	0.77	0.85	0.98	1.13	1.24
Oct. 26		0.19	0.29	0.38	0.43	0.45	0.46	0.48	0.49	0.49	0.50	0.53	0.55	0.56	

Source: Chow (8).

TABLE 4.4. Frequency analysis of exceedance values (10-min duration rainfall depth)

m	y (in.)	Tr (yrs)
(1)	(2)	(3)
1	1.11	35.00
2	0.96	17.50
3	0.94	11.67
4	0.92	8.75
5	0.88	7.09
6	0.80	5.833
7	0.80	5.000
8	0.76	4.375
9	0.74	3.839
10	0.74	3.500
11	0.71	3.182
12	0.70	2.917
13	0.68	2.692
14	0.68	2.500
15	0.68	2.333
16	0.67	2.188
17	0.66	2.039
18	0.66	1.944
19	0.66	1.812
20	0.65	1.750
21	0.64	1.667
22	0.64	1.591
23	0.63	1.522
24	0.62	1.458
25	0.62	1.400
26	0.61	1.346
27	0.60	1.296
28	0.60	1.250
29	0.59	1.207
30	0.59	1.167
31	0.58	1.129
32	0.58	1.094
33	0.57	1.061
34	0.57	1.029
35	0.57	1.000

Source: Chow (8).

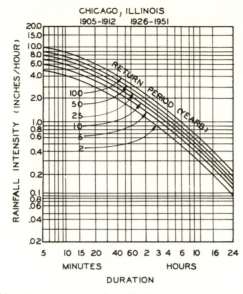

FIG. 4.7.   Intensity-duration-frequency curves (9).

to each rainfall T_r. This is a partial-duration series; therefore, the return period is given by the formula $T_r = N/m$, $N =$ years of record.

Next, the same type of analysis is carried out for a different duration, say 30 minutes. The 30-minute values may or may not include the 10-minute values of the preceding analysis. A frequency distribution is constructed from the 30-minute values, and the process is continued for other durations. The manner in which the precipitation data is reported has changed through the years, and modification of the records may be needed to put all the data on the same basis.

An analysis of the form indicated above can be used to produce an intensity-duration-frequency curve for Chicago, Ill., an example of which is shown in Figure 4.7.

This type of analysis was carried out on a nationwide basis and published by the Weather Bureau in a booklet of graphs from which Figure 4.7 was taken (9). Intensity-duration analysis can only be carried out in locations where recording rain gages are in operation. To facilitate general uses, the U.S. Weather Bureau has presented the results of their intensity-duration analyses in the form of a family of maps, some of which are reproduced in Figures 4.8–4.19 (10).

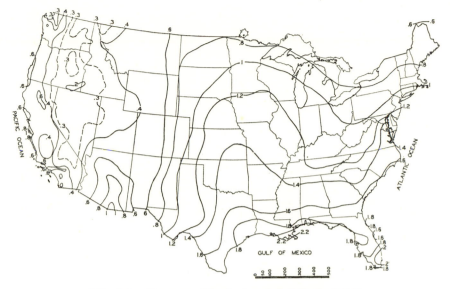

FIG. 4.8. Two-year 30-minute rainfall (inches) (10).

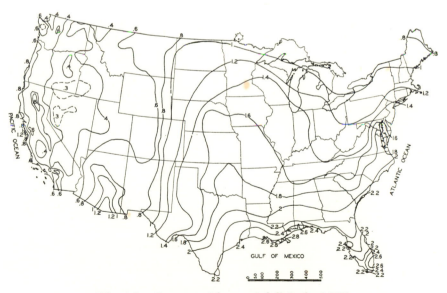

FIG. 4.9. Two-year 1-hour rainfall (inches) (10).

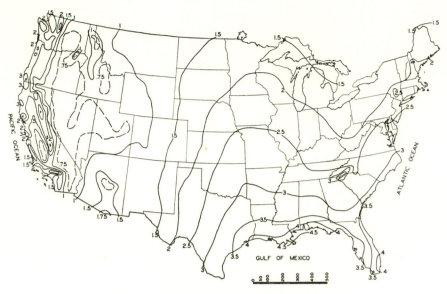

FIG. 4.10. Two-year 6-hour rainfall (inches) (10).

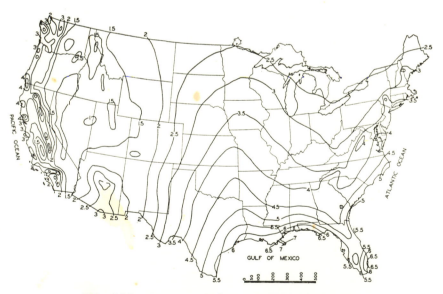

FIG. 4.11. Two-year 24-hour rainfall (inches) (10).

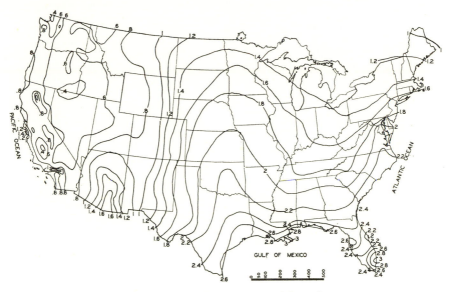

FIG. 4.12.   Ten-year 30-minute rainfall (inches) (10).

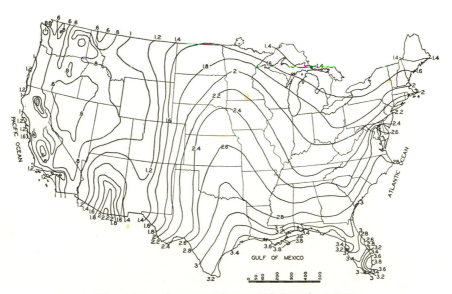

FIG. 4.13.   Ten-year 1-hour rainfall (inches) (10).

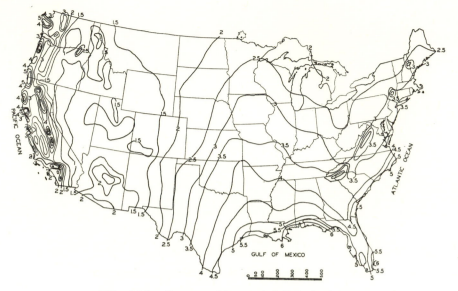

FIG. 4.14.   Ten-year 6-hour rainfall (inches) (10).

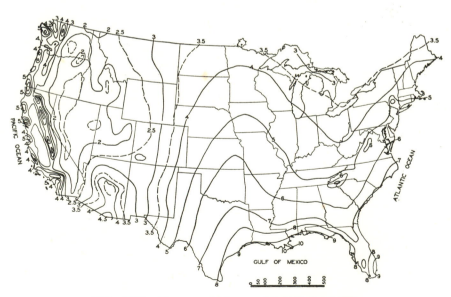

FIG. 4.15.   Ten-year 24-hour rainfall (inches) (10).

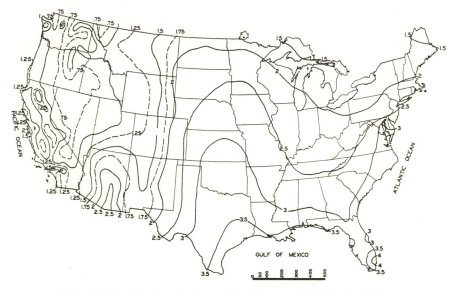

FIG. 4.16. One-hundred-year 30-minute rainfall (inches) (10).

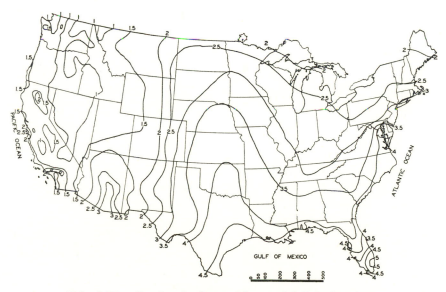

FIG. 4.17. One-hundred-year 1-hour rainfall (inches) (10).

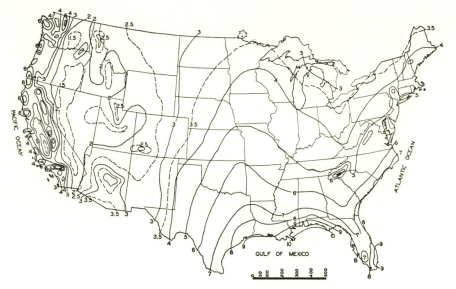

FIG. 4.18.   One-hundred-year 6-hour rainfall (inches) (10).

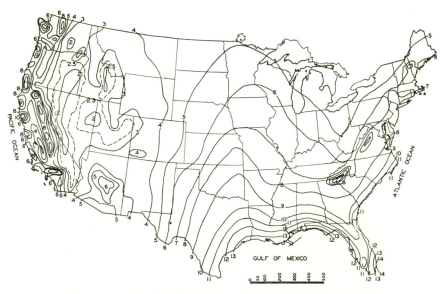

FIG. 4.19.   One-hundred-year 24-hour rainfall (inches) (10).

ANALYSIS OF PRECIPITATION OVER AN AREA

For the most part, engineering hydrologic studies require knowledge of precipitation over a definite area, the size of which may vary from a small parking lot to the drainage basin of a major stream. Because precipitation information collected with a rain gage represents conditions at a point, methods are required to transform point precipitation into information representative of the entire area.

STATION-AVERAGE METHOD

The simplest and most easily used method of estimating areal averages consists of simply computing the areal average of all observed values. This method is sound for a drainage basin having a relatively large number of uniformly distributed points at which precipitation has been measured. From Figure 4.20 the average precipitation would be computed as

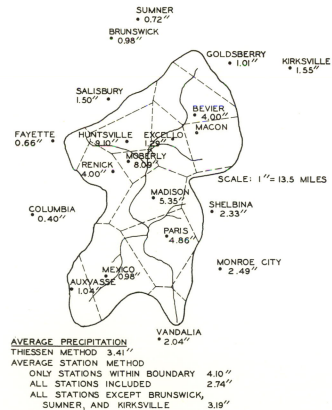

FIG. 4.20. Drainage basin used to illustrate computation of average precipitation.

$$P_{avg} = \left(\sum_{i=1}^{N} P_i \right) \Big/ N \qquad\qquad (4.1)$$

where

P_i = precipitation depth at station i
N = the total number of stations

This method assigns the same weight to each station regardless of location or other considerations. In Figure 4.20 stations outside the drainage area are also indicated, and the effect of including or excluding them in average precipitation calculations is indicated.

THIESSEN POLYGON METHOD
The Thiessen method weights each station in direct proportion to the area it represents without consideration of topography or other characteristics. The area represented by each station is assumed to be that which is closer to it than to any other station. In order to apportion the drainage area to the proper station, lines are first drawn connecting adjoining stations (shown as dotted lines in Fig. 4.20). Perpendicular bisectors are then constructed for each line joining two stations (shown as solid lines in Fig. 4.20). A set of perpendicular bisectors (with due allowance for the boundary of the drainage area) then encloses that part of the drainage area closest to the respective station. The average precipitation is determined in accordance with the equation

$$P_{avg} = \frac{\displaystyle\sum_{i=1}^{N} P_i \cdot A_i}{\displaystyle\sum_{i=1}^{N} A_i} \qquad\qquad (4.2)$$

where A_i is the area assigned to station i.
The Thiessen polygon method can be superior to the station average method because it accounts for nonuniform distribution of stations, and those not having an influence on the computation are automatically excluded.

ISOHYETAL METHOD
The isohyetal method provides a means of considering orographic or other effects in the computation of average precipitation. Contours of equal precipitation depth are constructed as illustrated in Figure 4.21.

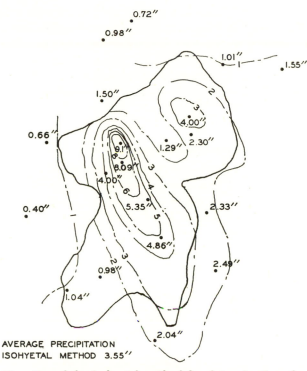

AVERAGE PRECIPITATION
ISOHYETAL METHOD 3.55"

FIG. 4.21.   Use of the isohyetal method for determination of average precipitation.

When contours are located, the effect of a mountain or a large hill can be taken into account as well as the direction of the storm's movement. Average precipitation between contours is assumed to be the numerical average of the two contour values. Thus the average precipitation over the entire area is computed by the equation

$$P_{\mathrm{avg}} = \frac{\sum\limits_{j=1}^{m} [(P_j + P_{j+1})(A_j/2)]}{\sum\limits_{j=1}^{m} A_j} \tag{4.3}$$

where

P_j = the precipitation at contour j
A_j = the area included between precipitation contours j and $j + 1$
m = the total number of different contours

The isohyetal method (isohyets are lines of equal precipitation depth) should be used when it is known that some storm distribution data are not reflected entirely by the station measurements. It is well to note that the Thiessen polygon method and the isohyetal method will yield the same value for average precipitation if the isohyets are drawn by linearly interpolating precipitation values along lines connecting adjacent stations.

CONSTRUCTION OF SYNTHETIC STORMS

Precipitation records are available for many more locations than are streamflow records. Frequently, however, analysis of an actual storm is difficult because of the time required to search records for the particular storm desired. Such a situation arises when a storm of particular duration and frequency is of interest. In that case it is convenient to use the results of the duration-frequency study made for the United States and to construct a synthetic storm (10). The procedure is as follows:

1. The return period is selected. If the return period does not correspond to one of those given on the maps (Figs. 4.8–4.19), a more extensive series of maps (10) should be consulted; or Table 4.5 as given by Davis and Sorenson (11) may be used.
2. Rainfall depths are read from the maps. An example is given in Table 4.6, column 2, for a point at the common corner of Illinois, Iowa, and Missouri. If the rainfall for durations other than those given is needed, the values from the maps should be graphed as shown in Figure 4.22 (10) and the desired values read from the graph. Values of accumulated rainfall for 2-hour increments are given in column 2 of Table 4.6.
3. The accumulated rainfall of column 2 is not in the sequence in which rainfall is usually accumulated. Recall that the 1-hour value

TABLE 4.5.   **Frequency of 6-hour rainfalls**

Return period (yrs)	Depth (% of 10-yr depth)
1	53
2	64
5	85
10	100
25	120
50	135
100	148

Source: Davis and Sorenson (11).

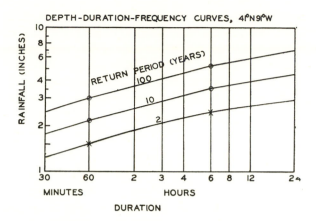

FIG. 4.22. Interpolation of depth-duration-frequency value (10).

TABLE 4.6. Construction of 100-year storm for a 100-square-mile drainage area (the common corner of Illinois, Iowa, and Missouri)

Time (hrs)	Accumulated rainfall (in.)	Increment (in.)	Rainfall sequence (in.)	Rainfall sequence average over area (in.)
(1)	(2)	(3)	(4)	(5)
0	0		0	
		3.9		
2	3.9		0.10	0.09
		0.8		
4	4.7		0.20	0.19
		0.4		
6	5.1		0.30	0.28
		0.3		
8	5.4		0.40	0.37
		0.3		
10	5.7		3.90	3.61
		0.2		
12	5.9		0.80	0.74
		0.2		
14	6.1		0.30	0.28
		0.2		
16	6.3		0.20	0.19
		0.15		
18	6.45		0.20	0.19
		0.15		
20	6.6		0.15	0.14
		0.10		
22	6.7		0.15	0.14
		0.10		
24	6.8		0.10	0.09

taken from the map is the catch for the most intense hour and thus also contains the most intense 30-minute period. In order to attain the proper sequence, the incremental rainfall is determined and is shown in column 3 of Table 4.6. The increments are rearranged in column 4 to provide a more realistic storm sequence. The most intense portion of the storm is located between the third point and the midpoint of the storm, and the other increments are rearranged to form an appropriate sequence, the strict arrangement of which is somewhat arbitrary.

Alternative approaches are available for establishing the storm sequence. The U.S. Soil Conservation Service, for example, suggests reading the 24-hour precipitation data from Figures 4.10, 4.14, and 4.18 and distributing this rainfall as indicated in Table 4.7 (12). Type 1 is applicable to Hawaii, Alaska, and the coastal side of the Sierra Nevada and the Cascade Mountains in California, Oregon, and Washington; Type 2 is applicable to the rest of the United States, Puerto Rico, and the Virgin Islands.

TABLE 4.7. Accumulation of rainfall to 24 hours

Time (hrs)	P_x/P_{24}*	
	Type 1	Type 2
0	0	0
2.0	0.035	0.022
4.0	0.076	0.048
6.0	0.125	0.080
7.0	0.156	. . .
8.0	0.194	0.120
8.5	0.219	. . .
9.0	0.254	0.147
9.5	0.303	0.163
9.75	0.362	. . .
10.0	0.515	0.181
10.5	0.583	0.204
11.0	0.624	0.235
11.5	0.654	0.283
11.75	. . .	0.387
12.0	0.682	0.663
12.5	. . .	0.735
13.0	0.727	0.772
13.5	. . .	0.799
14.0	0.767	0.820
16.0	0.830	0.880
20.0	0.926	0.952
24.0	1.000	1.000

Source: U.S. Soil Conservation Service (12).
* Ratio of accumulated rainfall P_x to 24-hour value P_{24}.

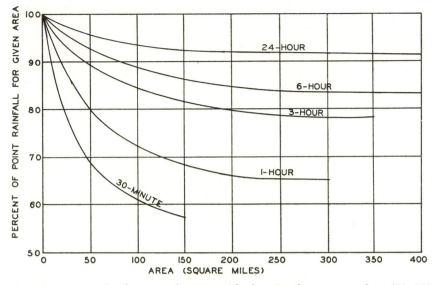

FIG. 4.23. Area-depth curves for use with duration-frequency values (10, 13).

The rainfall values shown in Figures 4.8–4.19 are point values. Thus, in the example, the synthesized storm is for a point in the watershed. This rainfall must be modified if it is to represent the conditions on the total watershed. How large an area is represented by the 8-inch diameter rain gage? The graph in Figure 4.23 is widely used as a guide to the variations between point and areal values of precipitation (10, 13). Column 5 of Table 4.6 has been computed by multiplying the values in column 4 by the appropriate percentage (0.93) chosen from Figure 4.23 for a drainage area of 100 square miles and a storm duration of 24 hours. Note that as the storm duration is greater, the point value of rainfall is more nearly equal to the average value.

PROBABLE MAXIMUM PRECIPITATION

Considering the probability of occurrence of extreme events, we might ask, How large is the largest possible flood? Probabilistic analysis alone could indicate that as the probability of occurrence approaches zero, the magnitude of the flood becomes infinitely large. However, because floods result directly or indirectly from precipitation, it is logical to assume that physical limitations do establish a finite maximum size.

Some feeling about the impact of a storm as intense as the probable

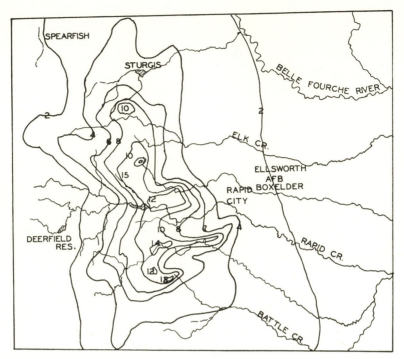

FIG. 4.24. Isohyetal map of the storm at Rapid City, S. Dak., June 9–10, 1972 (14).

maximum arises from consideration of the effects of the severe storm experienced by Rapid City, S. Dak., on June 9, 1972. Figure 4.24 is an isohyetal map of that storm (14). Up to 15 inches of precipitation occurred within 6 hours. Usually placid, Rapid Creek flowed at 50,600 cfs as a result of the storm. Some $100 million in damage occurred and 236 people lost their lives in the flood. Computations of probable maximum precipitation (PMP) for the Rapid City area indicate 20 inches in 6 hours. Thus the terrible flood was the result of a storm which probably did not even equal the PMP value.

On occasion dams or other structures are built at locations where failure could produce catastrophic damage. For such an event humanitarian (and perhaps economic) considerations make it desirable to design the project to be safe in the event of the occurrence of the largest probable flood. Since such a flood would conceivably occur as the result of the PMP, estimates of the realistic upper limits of precipitation have been made (15, 16).

These estimates of PMP have been formed from analysis of record historical storms, mathematical models of intense storms, and moisture available for precipitation (15, 16, 17). In general, storms of record are transposed from the area where they actually occurred to the region of interest. Transposition of the storms is subject to limitations. Topography of the areas must be similar as well as other general meteorological conditions. For example, a storm occurring in New Jersey could not be transposed to the mountains of California. Precipitation from the actual storm is adjusted according to the ratio of moisture available in a column of atmosphere at the time of the storm to the probable maximum that could be available.

Figure 4.25 is a compilation of record storms of various durations that have actually been recorded (18). Estimates of PMP correspond closely to many of these storms. Figure 4.26 (16) summarizes the estimates of PMP and can be used to construct synthetic storms in the same manner as Figures 4.8–4.19 were used.

The average 24-hour precipitation for a watershed of 200 square miles is read from Figure 4.26. Note that the United States is divided into several zones. The graphs of Figure 4.27 give factors for converting to different durations and different areas (16). Thus when the 24-hour value from Figure 4.26 and the factors from Figure 4.27 are used, the cumulative depths at 6, 12, 24, and 48 hours are established. The U.S. Bureau of Reclamation (19) recommends breaking up the first six hours according to the percentages given in Table 4.8. The cumulative distribution is then broken into rainfall increments, which are rearranged in the sequence: 6th hour, 4th hour, 3rd hour, 1st hour, 2nd hour, 5th hour.

TABLE 4.8. Six-hour probable maximum precipitation distribution

Time (hrs)	Depth (% of 6-hr depth)
1	49
2	64
3	75
4	84
5	92
6	100

Source: U.S. Bureau of Reclamation (19).

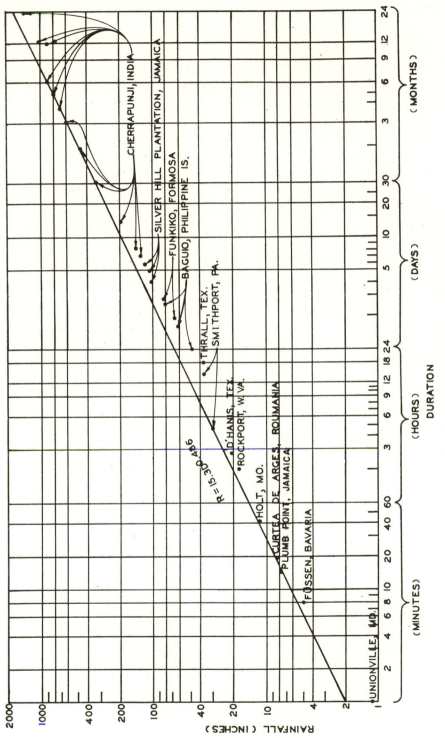

FIG. 4.25 The world's record point rainfall values (18).

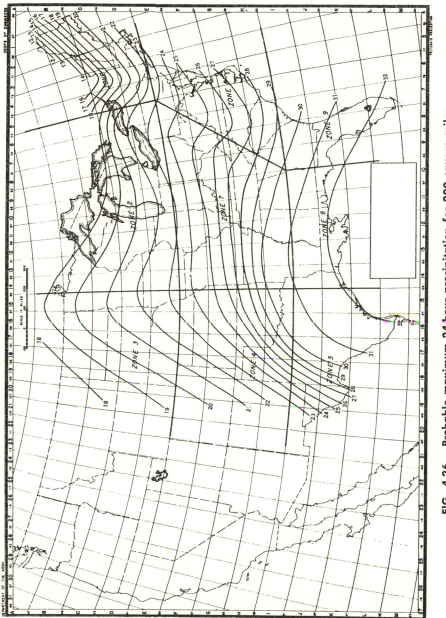

FIG. 4.26. Probable maximum 24-hour precipitation for 200 square miles (inches). The all-season envelope (16).

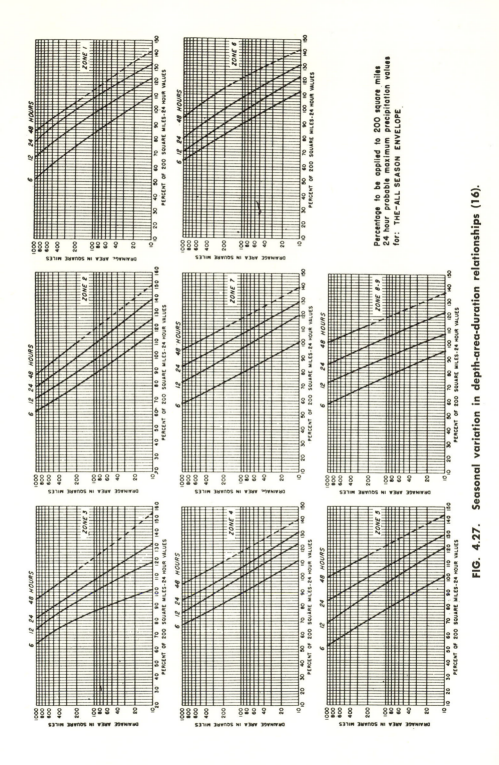

Percentage to be applied to 200 square miles
24 hour probable maximum precipitation values
for: THE-ALL SEASON ENVELOPE

FIG. 4.27. Seasonal variation in depth-area-duration relationships (16).

PROBLEMS

4.1. What is the depth of a 2-hour rainfall that would occur once in 10 years at Chicago, Ill.?

4.2. A 24-hour storm in Kansas City produced 7.6 inches of rainfall. Estimate the recurrence interval of this storm.

4.3. Construct a realistic 25-year 12-hour storm as it might have occurred for your hometown.

4.4. Construct a 50-year 24-hour storm giving average precipitation on a 200-square-mile area in the center of Kansas.

4.5. Using the station-average method, determine average precipitation for the given drainage basin using the given data on precipitation.

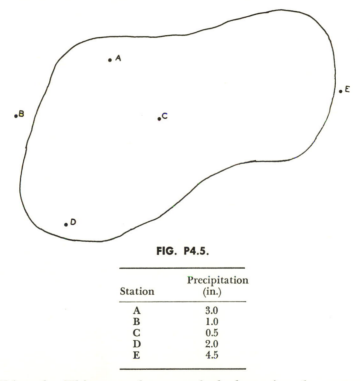

FIG. P4.5.

Station	Precipitation (in.)
A	3.0
B	1.0
C	0.5
D	2.0
E	4.5

4.6. Using the Thiessen polygon method, determine the average precipitation for the storm and drainage basin given in Problem 4.5.

4.7. Using an isohyetal map, determine the average precipitation for the storm and drainage basin given in Problem 4.5.

4.8. In Figures 4.8–4.19 why do the precipitation contours for California differ in form so much from those of the midwestern states?

REFERENCES

1. Perrault, Pierre. *Origin of Fountains* (trans. A. La Rogue). Hafner, 1967.
2. Biswas, Asit K. Beginning of quantitative hydrology. *ASCE J. Hydraul. Div.,* Sept. 1968.
3. U.S. Dept. of Agriculture. *Climate and Man.* Yearbook Agr., Washington, D.C., 1941.
4. Shames, I. H. *Engineering Mechanics.* Prentice-Hall, 1967.
5. Bruce, J. P.; and Clark, R. H. *Introduction to Hydrometeorology.* Pergamon Press, London, 1966.
6. U.S. Weather Bureau. Climatological data. Washington, D.C. (for daily records of rainfall). (Hourly rainfall intensities are found in hydrologic bulletin, 1940–48; climatological data, 1948–51; and hourly precipitation data, 1951–.)
7. U.S. Geological Survey. Double mass curves. Water Supply Paper 1941-B, Washington, D.C., 1960.
8. Chow, Ven Te. Frequency analysis of hydrologic data with special application to rainfall intensities. Univ. Ill. Eng. Exp. Sta. Bull. Ser. 414, 1953.
9. U.S. Weather Bureau. Rainfall intensity-duration-frequency curves. Tech. Paper 25, Washington, D.C., 1955.
10. ———. Rainfall frequency atlas of the United States. Tech. Paper 40, Washington, D.C., 1961.
11. Davis, V. C.; and Sorenson, K. E. *Handbook of Applied Hydraulics.* McGraw-Hill, 1969.
12. U.S. Soil Conservation Service. Watershed planning. *In* National Engineering Handbook, Sect. 4, Part 1, Washington, D.C., 1964.
13. U.S. Weather Bureau. Rainfall intensity-frequency regime. Tech. Paper 29, Washington, D.C., 1957.
14. Thompson, H. J. The Black Hills flood. *Weatherwise,* Aug. 1972.
15. U.S. Weather Bureau. Generalized estimates of maximum possible precipitation over the United States east of the 105th meridian. Hydrometeorological Rept. 23, Washington, D.C., June 1947.
16. ———. Generalized estimates of maximum possible precipitation for the United States west of the 105th meridian. Tech. Paper 38, Washington, D.C., 1960.
17. Wiesner, C. J. *Hydrometeorology.* Chapman and Hall, London, 1970.
18. Jennings, A. H. World's greatest observed point rainfall. *Monthly Weather Rev.,* Vol. 78, 1950.
19. U.S. Bureau of Reclamation. Design of small dams. Washington, D.C., 1965.

CHAPTER FIVE

INFILTRATION

IN THE PRECEDING CHAPTER precipitation data were analyzed and synthetic rainstorms produced. Water thus arrives at ground level. Some of this water will seep into the ground and some will run off. In this chapter the separation between that which runs off and that which infiltrates into the ground will be considered. Only the separation into the two quantities is obtained; the process of transporting the runoff quantity to the point of interest is covered in Chapter 6.

The infiltration process is very complex and only partially understood. The commonly used methods for estimating the infiltration quantities will be described in the following section.

INFILTRATION PROCESS

A falling drop of water may be intercepted by vegetation or may fall directly on the land. The quantity intercepted is difficult to estimate and for many purposes is considered as part of the infiltration. After the vegetation becomes wet, additional raindrops run off and drop to the ground.

Water that reaches the ground may evaporate back to the atmosphere, enter the ground, or run off. The rate at which water enters the ground is called the infiltration rate. Evaporation is greatly reduced during a storm due to the high humidity. Initial runoff is caught by the surface depressions, which must be filled before further runoff can make its way to a stream or river. Infiltration, however, starts when the first drop of water touches the ground surface and continues even after precipitation ceases until all depressions in the land surface are empty. The part of precipitation that eventually reaches a stream is called the rainfall excess.

The infiltration rate may be limited in two ways. It cannot exceed the rate at which water is added to the surface (i.e., by the rainfall intensity), and it is limited by the rate at which water can enter and move through the soil. Three separate processes make up infiltration: passage through the soil surface, movement through the soil, and depletion of the moisture retention capacity of the soil.

The ability of water to pass through the surface depends upon the physical condition of the soil and its covering vegetation. During dry periods the soil and the vegetal root systems shrink, opening passages for easy entry of water into the soil. Moisture causes both soil and vegetation to swell, closing these passages. Flowing water washes fine material into the soil pores, clogging the openings, and the impact of falling raindrops compacts the soil surface. As a result, during a storm the potential infiltration rate starts at a maximum and decreases as time progresses.

Transmission through the various soil strata depends upon the mechanical composition of the soil, the primary factor being the permeability or size and quantity of interconnected pores in the soil. The soil pores can store water to a certain upper limit; the soil thus acts as a sponge. Initially, the moisture entering the sponge is stored there; when the storage capacity is reached, no further storage is possible and the water passes through. The soil acts in a similar fashion with the infiltration rate decreasing as the soil storage capacity is depleted.

Type and extent of covering vegetation also affect the potential infiltration rate by breaking the fall of the raindrops. This also influences the rate at which runoff crosses the land and thus affects the total infiltration quantity.

Robert Horton (1) proposed representing the infiltration rate by an exponentially decaying formula of the form

$$f = f_c + (f_0 - f_c)e^{-k_f t} \tag{5.1}$$

where

f = infiltration rate at time t
f_0 = initial infiltration rate at time $t = 0$
f_c = infiltration rate which f approaches asymptotically
k_f = a constant
t = time

Equation (5.1) is widely accepted as a reasonable representation of the infiltration process. Problems associated with determining the con-

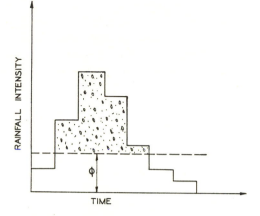

FIG 5.1. Determination of the ϕ index.

stants f_c, f_o, and k_f have precluded wide use of the equation in practice. Approximate methods and averages are usually used, some of which are discussed in the following sections.

INFILTRATION INDICES

Many indices have been proposed as indicators of infiltration. The simplest and most widely used is the ϕ index. This index is defined as the average rainfall intensity above which the mass of runoff is equal to the mass of rainfall. Thus ϕ has the units of rainfall intensity, inches/hour. The ϕ index is determined from measured rainfall and runoff for a particular watershed.

The time distribution of average rainfall depth is determined by applying the Theissen method or other approach to hourly rainfall values. The result is indicated as the solid line in Figure 5.1. Volume of runoff is determined by using methods described in Chapter 6. The difference between the rainfall and runoff volumes is considered to be the infiltration volume.

Thus to determine the ϕ index, a value of ϕ (in inches/hour) is assumed (the dashed line in Fig. 5.1). This assumed value of ϕ is subtracted from the hourly rainfall values. Negative values indicate rainfall intensity less than infiltration rate and are set to zero. The difference computed is the rainfall excess. The volume of rainfall excess must equal the volume of runoff as measured if the correct value of ϕ was assumed. If not, a new value of ϕ is assumed and the process is repeated.

The ϕ index includes the infiltration quantity, the intercepted quantity, and the quantity left in depression storage. A portion of the latter quantity will evaporate instead of infiltrating.

In an attempt to separate noninfiltration quantities from the ϕ index, the f_{avg} index was introduced. The f_{avg} represents the average infiltration rate for the duration of the infiltration period and may be taken as

$$f_{avg} = (1/T) \int_0^T [f_c + (f_0 - f_c)e^{-k_f t}] \, dt \qquad (5.2)$$

if the rainfall intensity exceeds the infiltration rate f, as given by Eq. (5.1), throughout the duration of the storm T.

The parameters f_c, f_0, and k_f depend upon the soil-cover complex. If these quantities vary across the watershed, the f_{avg} is an average quantity for the watershed and is termed the W index. Because f_{avg} and W are related to Eq. (5.1), the disadvantages of Eq. (5.1) also apply to Eq. (5.2). As a result, the ϕ index is the most easily used.

ESTIMATION OF RAINFALL EXCESS
BY SOIL-COVER COMPLEX ANALYSIS

The work of the U.S. Soil Conservation Service requires determination of infiltration quantities for watersheds on which no measurements of rainfall or runoff are available. Thus determination of any of the standard indices is very difficult. Detailed information is available, however, on the soil and on the cover. Thus a process was developed for determining the rainfall excess based on the soil-cover complex.

The following description of the principles on which the procedure was developed is, with slight modification, from Section 4 of the SCS National Engineering Handbook:

If records of natural precipitation and rainfall excess for a large storm over a small area are used, a plotting of accumulated rainfall excess versus accumulated rainfall will show that rainfall excess starts after some rain accumulates (there is an "initial abstraction" of rainfall) and that the line curves, becoming asymptotic to a straight line. On arithmetic graph paper and with equal scales the straight line has a 45° slope. The relation between precipitation and rainfall excess can be developed from this plotting, but a better understanding of the relation is had by first studying a storm in which precipitation and rainfall excess begin simultaneously (the initial abstraction does not occur). For the simpler storm the relation between precipitation, rainfall excess, and retention (the rain not converted to runoff) at any point on the mass curve can be expressed as

$$F/S' = p/P \qquad (5.3)$$

where

F = actual retention
S' = potential maximum retention $(S' \geqq F)$
p = actual rainfall excess
P = potential maximum rainfall excess $(P \geqq p)$

The parameter S' in Eq. (5.3) does not contain the initial abstraction and differs from the parameter S to be used later. The retention S' is a constant for a particular storm because it is the maximum that can occur under the existing conditions if the storm continues without limit. The retention F varies because it is the difference between P and p at any point on the mass curve, or

$$F = P - p \qquad (5.4)$$

Equation (5.3) can therefore be rewritten as

$$(P - p)/S' = p/P \qquad (5.5)$$

Solving for p produces

$$p = P^2/(P + S') \qquad (5.6)$$

which is a precipitation–rainfall excess relation in which the initial abstraction is ignored. The initial abstraction is brought into the relation by subtracting it from the rainfall. The equivalent of Eq. (5.3) becomes

$$F/S = p/(P - I_a) \qquad (5.7)$$

where I_a is the initial abstraction, $F \leqq S$, and $p \leqq (P - I_a)$. The parameter S includes I_a; i.e., $S = S' + I_a$. Equation (5.4) becomes

$$F = (P - I_a) - p \qquad (5.8)$$

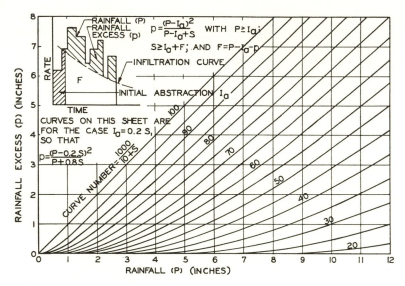

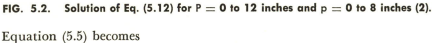

FIG. 5.2. Solution of Eq. (5.12) for P = 0 to 12 inches and p = 0 to 8 inches (2).

Equation (5.5) becomes

$$[(P - I_a) - p]/S = p/(P - I_a) \tag{5.9}$$

and Eq. (5.6) becomes

$$p = (P - I_a)^2/[(P - I_a) + S] \tag{5.10}$$

which is the precipitation–rainfall excess relation with the initial abstraction taken into account. The initial abstraction consists mainly of interception, infiltration, and surface storage, all of which occur early in the process before runoff reaches a stream. The insert on Figure 5.2 shows the position of I_a in a typical storm. To remove the necessity for estimating these variables in Eq. (5.10), the relation between I_a and S (which includes I_a) was developed by means of rainfall and runoff data from experimental small watersheds. The empirical relationship is

$$I_a = 0.2S \tag{5.11}$$

Substituting Eq. (5.11) into Eq. (5.10) gives

$$p = (P - 0.2S)^2/(P + 0.8S) \tag{5.12}$$

which is the precipitation–rainfall excess relation used in the SCS method of estimating direct runoff from storm rainfall (2).

TABLE 5.1. Soil groupings

Group	Minimum infiltration rate (in./hr)	Soils
A	0.30–0.45	deep sand, deep loess, aggregated silts
B	0.15–0.30	shallow loess, sandy loam
C	0.05–0.15	clay loams, shallow sandy loam, soils low in organic content, and soils usually high in clay
D	0–0.05	soils which swell significantly when wet, heavy plastic clays, and certain saline soils

Source: U.S. Soil Conservation Service (2).

At this point it is obvious that the parameter S is unknown and must be established from the soil-cover complex. The soil alone is classified initially in the form shown in Table 5.1. A detailed list in which named soil types are associated with the groups in the table is given in the SCS National Engineering Handbook (2).

It is necessary to include the soil cover in the description of the soil-cover complex. This is done through Table 5.2 (2). Using the soil group and the cover existing in the drainage basin, we read a curve number CN. This curve number is related to S, which was to be determined by the equation

$$CN = 1000/(10 + S) \qquad (5.13)$$

The hydrologic condition of the soil-cover complex must be estimated. As an example consider pasture or range:

Poor—Heavily grazed, no mulch or less than $\frac{1}{2}$ of area with plant cover.
Fair—Moderately grazed. Has plant cover on $\frac{1}{2}$ to $\frac{3}{4}$ of area.
Good—Lightly grazed. Has plant cover on more than $\frac{3}{4}$ of area.

Obviously, management practice is an index of the hydrologic condition, especially for row crops and small grains.

The discussion of the exponential decay curve for infiltration indicates that the antecedent moisture conditions have considerable influence on infiltration in the initial portion of the storm. In order to estimate this influence, the following antecedent moisture conditions are defined:

Condition I—Soils are dry but not to the wilting point, and satisfactory cultivation has taken place.

TABLE 5.2. Curve numbers for hydrologic soil-cover complexes, antecedent moisture condition II, $I_a = 0.2S$

Land use	Cover Treatment or practice	Hydrologic condition	Hydrologic soil group A	B	C	D
Fallow	straight row	. . .	77	86	91	94
Row crops	straight row	poor	72	81	88	91
	straight row	good	67	78	85	89
	contoured	poor	70	79	84	88
	contoured	good	65	75	82	86
	contoured and terraced	poor	66	74	80	82
	contoured and terraced	good	62	71	78	81
Small	straight row	poor	65	76	84	88
grain	straight row	good	63	75	83	87
	contoured	poor	63	74	82	85
	contoured	good	61	73	81	84
	contoured and terraced	poor	61	72	79	82
	contoured and terraced	good	59	70	78	81
Close-seeded	straight row	poor	66	77	85	89
legumes*	straight row	good	58	72	81	85
or	contoured	poor	64	75	83	85
rotation	contoured	good	55	69	78	83
meadow	contoured and terraced	poor	63	73	80	83
	contoured and terraced	good	51	67	76	80
Pasture		poor	68	79	86	89
or range		fair	49	69	79	84
		good	39	61	74	80
	contoured	poor	47	67	81	88
	contoured	fair	25	59	75	83
	contoured	good	6	35	70	79
Meadow		good	30	58	71	78
Woods		poor	45	66	77	83
		fair	36	60	73	79
		good	25	55	70	77
Farmsteads		. . .	59	74	82	86
Roads						
dirt†		. . .	72	82	87	89
hard surface†		. . .	74	84	90	92

Source: U.S. Soil Conservation Service (2).
* Close-drilled or broadcast-seeded.
† Including right-of-way.

Condition II—Average case for annual floods, i.e., an average of conditions that have preceded the occurrence of maximum annual floods on numerous watersheds.

Condition III—Heavy rainfall or light rainfall and low temperatures have occurred during the five days preceding the storm, and the soil is nearly saturated.

**TABLE 5.3.   Estimation of curve number for various anteced-
ent moisture conditions**

Curve number for condition	Corresponding curve number for condition	
II	I	III
100	100	100
95	87	99
90	78	98
85	70	97
80	63	94
75	57	91
65	45	83
60	40	79
55	35	75
50	31	70
45	27	65
40	23	60
35	19	55
30	15	50
25	12	45
20	9	39
15	7	33
10	4	26
5	2	17
0	0	0

Source: U.S. Soil Conservation Service (2).

Table 5.2 gives curve numbers for condition II. To convert these curve numbers to the other antecedent moisture conditions, Table 5.3 is used (2).

An estimate of the rainfall excess may now be determined from the total precipitation. The steps are as follows:

1. Tabulate accumulated precipitation at corresponding times (see Table 5.4). In the example, column 1 is the time and column 2 is the precipitation sequence and is identical to column 5 of Table 4.6 in which the storm was generated. Column 3 is the accumulated precipitation depth for that storm.
2. Estimate the accumulated rainfall excess at each time using the *CN*, the accumulated rainfall of column 3, and either Figure 5.2 or 5.3 (2). The accumulated rainfall excess is tabulated in column 4.
3. Compute the increments of rainfall excess. The increments are the differences between successive values in column 4 and are tabulated in column 5. The increments will be used later to establish the streamflow hydrograph.

TABLE 5.4. Example rainfall excess computation

		(Curve number = 80) (Antecedent moisture condition = II)		
Time (hrs)	Rainfall sequence (in.)	Accumulated rainfall depth (in.)	Accumulated rainfall excess (in.)	Rainfall excess increments (in.)
(1)	(2)	(3)	(4)	(5)
0		0	0	
	0.09			
2		0.09	0	
	0.19			
4		0.28	0	
	0.28			
6		0.56	0	
	0.37			0.09
8		0.93	0.09	
	3.61			2.41
10		4.54	2.50	
	0.74			0.60
12		5.28	3.10	
	0.28			0.30
14		5.56	3.40	
	0.19			0.15
16		5.75	3.55	
	0.19			0.15
18		5.94	3.70	
	0.14			0.15
20		6.08	3.85	
	0.14			0.15
22		6.22	4.00	
	0.09			0.10
24		6.31	4.10	

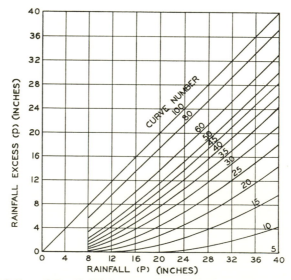

FIG. 5.3. Solution of Eq. (5.12) for P = 8 to 40 inches and p = 0 to 40 inches (2).

The U.S. Soil Conservation Service method is applicable to any size watershed. If, however, the watershed varies in soil type or in cover, the watershed should be broken into regions of similar character and the regions analyzed separately. It should also be noted that very large watersheds seldom experience uniform rainfall intensity throughout their extent.

PROBLEMS

5.1. The average rainfall over a 120-acre watershed for a particular storm was determined to be

Hour	Hourly rainfall (in.)
0	0
1	0.2
2	0.4
3	1.5
4	1.0
5	0.5
6	0.2
7	0

The volume of runoff from this storm was determined to be 22 acre-feet. What is the ϕ index?

5.2. Using the SCS method, determine the runoff from a 12-hour, 50-year storm and a probable maximum storm for a watershed in your vicinity or for one of the watersheds given in the Appendices. Make appropriate assumptions with regard to soil, usage, and condition.

REFERENCES

1. Horton, R. E. The role of infiltration in the hydrologic cycle. Trans. Am. Geophys. Union, Vol. 14, pp. 446–60, 1933.
2. U.S. Soil Conservation Service. Watershed planning. *In* National Engineering Handbook, Sect. 4, Part 1, Washington, D.C., 1964.

CHAPTER SIX
STREAMFLOW

NEARLY ALL LAND DEVELOPMENT PROJECTS involve drainage in some manner. Runoff from rainstorms must be drained from roadways in order to prevent hazardous driving conditions. Bridges and their substructures must be designed to pass a reasonable flood without objectionable scour at abutments or bridge piers or other damage to the bridge. Dams, spillways, airports, parking lots, shopping centers, flood plains, and city streets all require consideration of runoff for proper engineering design. The time distribution of this runoff is called the hydrograph. In the hydrologic cycle it is the streamflow. The engineer or designer is most frequently interested in obtaining the streamflow hydrograph for a particular storm. The specifications of the storm he picks can be chosen by the methods of Chapter 3.

THE HYDROGRAPH

A hydrograph is a graph of discharge passing a particular point on a stream, plotted as a function of time. Because discharge for a particular channel can be expressed in terms of depth of flow or stage, the hydrograph is frequently plotted as stage versus time. In fact, at a gaging station where discharge is measured, stage is measured and/or recorded.

As noted in the hydrologic cycle, streamflow is made up of rain falling directly on the stream, surface runoff, and flow from groundwater. Surface runoff reaches the stream by the process of overland flow. Rain falling directly on the stream produces streamflow almost instantly. Because groundwater flow moves slowly compared to overland flow, infiltration produces streamflow much more slowly than direct rainfall. Surface runoff, in general, produces the largest and most rapidly ap-

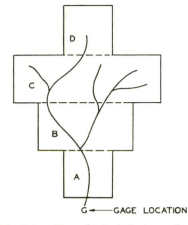

G ◄──────GAGE LOCATION

FIG. 6.1. Hypothetical drainage basin.

parent contribution to streamflow and thus is most instrumental in producing the peak flows.

Chapter 5 dealt with the estimation of rainfall excess, given a storm
of particular magnitude and duration. This chapter is primarily devoted to obtaining the streamflow hydrograph, given the time distribution of rainfall excess.

An understanding of the runoff process is gained by considering the
following example. A watershed is shown in Figure 6.1. The watershed
has been arbitrarily divided into four sections (A, B, C, D), with the area
of each given in Table 6.1. The runoff from each section will arrive at
the gaging station G at different times. These times are also listed in
Table 6.1, assuming that the storm begins at time zero.

Consider a rainfall intensity of 1 inch/hour covering the basin uniformly and falling continuously for 5 hours. At the end of the first hour
7 acre-inches of rainfall are on the basin. If we assume no infiltration,
all of this runs off; only that falling on section A reaches the gage G by
the end of this first hour. The remainder is still on the land or in the
channel traveling toward G. The 2 acre-inches that fell on B will not
reach G for another hour, that which fell on C will not reach G for 3

**TABLE 6.1. Time of flow of surface runoff from subsections to gage lo-
cation for a hypothetical watershed**

Item	Subsection			
	A	B	C	D
Area (acres)	1	2	3	1
Time to G (hrs)	1	2	3	4

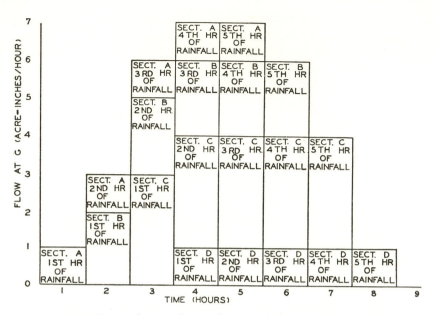

FIG. 6.2. Runoff contributions from subareas of the watershed in Figure 6.1.

hours, and so on. This process is indicated by the blocks drawn in Figure 6.2. Each block is designated by the section from which it originated and by the rainfall increment from which it was generated.

This block diagram points out significant features of hydrographs: the rainfall lasted only 5 hours, but the runoff lasted for 8 hours; the duration of the storm is not the same as the time base of the hydrograph.

The hydrograph in Figure 6.2 is schematic. All parts of each subarea would not begin contributing to the hydrograph at the same time as is implied by the use of blocks to represent runoff. If the drainage area could be divided into a very large number of subareas, a more continuous hydrograph would be obtained that would more nearly represent actual conditions.

It took 4 hours before all portions of the watershed were contributing to the flow at point G, the condition producing peak flow. If the storm had lasted less than 4 hours, not all portions would have contributed simultaneously; but because the rainfall was of constant intensity and lasted longer than 4 hours, a flat peak resulted. The time required for all portions of the watershed to contribute to the flow at G is quite important and is called the time of concentration.

The example depicts the flow resulting from storm runoff. This flow is superimposed onto the baseflow, the flow that would have oc-

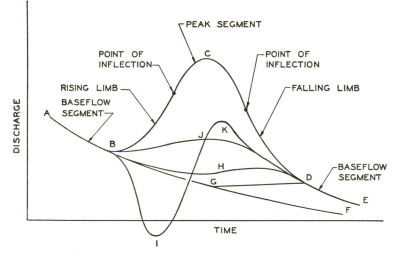

FIG. 6.3. Baseflow separation.

curred had there been no surface runoff. For discussion purposes the hydrograph is divided into the baseflow segment, the rising limb, the peak segment, and the falling limb. These segments are indicated in Figure 6.3.

BASEFLOW RECESSION CURVES

In previous chapters storms with given return periods and maximum probable storms were discussed. The infiltration component was extracted from these storms to yield the resulting surface runoff (rainfall excess). This quantity of runoff was left on the land. The time of conveyance, or routing, of this flow to the point of interest (overland flow and streamflow) remains. This will be accomplished by attaining a representative runoff hydrograph. In order to derive the runoff hydrograph, the baseflow (that part contributed by groundwater) must be subtracted from a measured hydrograph.

The material covered in Chapter 1 relative to the hydrologic cycle and the response of a watershed indicated that the baseflow would vary with the moisture condition in the root zone and with the intensity and duration of the storm. This variation is understood qualitatively but its quantitative estimate is crude. It is as yet impossible to accurately analyze the effects of all parameters influencing infiltration and groundwater contribution.

Several possible conditions are indicated in Figure 6.3 in which the

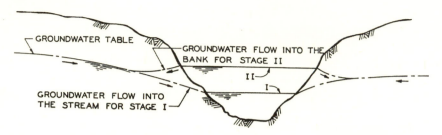

Fig. 6.4. Groundwater behavior during baseflow (I) and during flooding (II).

measured hydrograph is *ABCDE*. Line *BGF* is an extension of line *AB,* which represents the groundwater contribution if rainfall had not occurred. If no increase in groundwater occurs through infiltration, the hydrograph should return to this line when runoff stops. This is not the case in this example. Line *DE* is the baseflow occurring after the end of the runoff, and it has been displaced vertically upward from the baseflow prior to the storm. This is another baseflow recession curve but should have the same shape as *ABGF* if, as is usually assumed, the baseflow curve has a constant shape.

In Figure 6.3 the later recession curve has been extrapolated backward to point *K*. Points *B* and *D,* where the baseflow curves deviate from the hydrograph, indicate respectively the beginning and ending of runoff. How do these two points connect? It is instructive to consider what happens to the baseflow during a storm. Figure 6.4 represents a cross section of a stream and adjoining land area. Stream water surface I represents a baseflow condition during which groundwater flows from the bank to the stream. Consider now a storm occurring so that the stage of the stream rises to level II. The hydrostatic pressure on the bank due to streamflow is now greater than that in the water table, and groundwater flow actually reverses, with water flowing from the stream into the bank. In Figure 6.3 this condition is actually a negative discharge since the process subtracts from streamflow. The water entering the bank is called bank storage. In rivers such as the Missouri where the hydrograph rises slowly and recedes slowly, bank storage can be significant (1).

Line *BID* in Figure 6.3 qualitatively represents the manner in which groundwater flow should behave during a period of rise in the stream. As the stage rises, the flow of groundwater decreases and eventually goes negative. If the soil in the drainage basin is highly permeable so that the groundwater table rises more rapidly than the stream stage, the baseflow may be represented by a curve such as *BJD*. Often a straight line from *B* to *D* or from *G* (a point under the hydrograph peak) to *D* is drawn and is assumed to represent the division between groundwater flow and surface runoff. The area below this assumed curve and the area between the baseflow curve and the hydrograph represent groundwater contribution and surface runoff respectively.

No matter which of the assumed baseflow curves shown in Figure 6.3 is used, the total surface runoff volume (area above the baseflow curve) is approximately the same. The time distribution is different, however, in each case. The difference between the ordinates of the baseflow curve and the hydrograph is the flow rate of surface runoff; therefore, curve *BID* indicates that the surface runoff peak occurs earlier than indicated by *BJD, BD,* or *BGD*.

In order to estimate the baseflow, curves *AB* and *DE* must be extended to *G* and *K* respectively. The baseflow curve is frequently assumed to be a decreasing function of the form (2)

$$q_{t+\Delta t} = q_t K^{\Delta t}; \quad K < 1.0 \tag{6.1}$$

in which

$$\Delta t = \text{an increment in time } t$$
$$q_t = \text{flow at time } t$$
$$q_{t+\Delta t} = \text{flow at time } t + \Delta t$$
$$K = \text{a constant to be determined}$$

Equation 6.1 implies that groundwater flows from a reservoir whose discharge is linearly proportional to storage. For a particular drainage basin, K is determined from the analysis of a known hydrograph. If Δt is taken to be one time unit,

$$q_{t+\Delta t} = q_t K \tag{6.2}$$

Note that the value of K is dependent upon the magnitude chosen for Δt; $q_{t+\Delta t}$ is plotted against q_t as shown in Figure 6.5. Only values representing flow recession are used. If the relationship implied in Eq. (6.1) were exact, the points representing the recession portions of the hydrograph would plot as a straight line. This recession region is necessarily the portion of Figure 6.5 for low q values. The line enveloping the data at the top is taken as the baseflow recession and indicates that $K = 0.97$. The more rapid drop in flow from one day to the next indicated by the lower curve is taken as an indication of the direct-runoff recession, or the falling-limb portion of the runoff hydrograph.

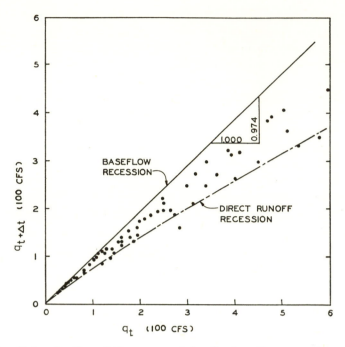

FIG. 6.5. Determination of the exponent for the baseflow recession equation.

An added aspect of Eq. (6.1) is that if the logarithm of each side is taken, one obtains

$$\log q_{t+\Delta t} = \log q_t + {}^{\Delta t} \log K \qquad (6.3)$$

This indicates that if K is constant and Eq. (6.3) is plotted on semilogarithmic graph paper with q_t on the log scale and time as the arithmetic scale, a straight line will result. Similarly, a straight line will also result for the direct-runoff recession. This has led to use of semilog plots for hydrograph separation (3). Once K has been computed, the baseflow curve can be extended ahead from A and backward from E (Fig. 6.3) using Eq. (6.1). The points at which the extension deviates from the baseflow curve (B and D) are located, and the proper curve from B to D can be sketched.

Total runoff volume is obtained by integration of the area above the baseflow curve. Integration is usually accomplished with a planimeter or by numerical methods. Commonly used units of volume are cfs-days or cfs-hours.

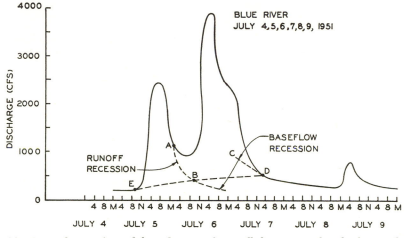

FIG. 6.6.   Separation of baseflow and runoff for a complex hydrograph.

COMPLEX HYDROGRAPHS

It is not always possible to obtain desired hydrographs resulting from an isolated storm, and frequently one influenced by two separate storms must be considered. Figure 6.6 illustrates such a complex hydrograph. In this case it is necessary to separate the total flow according to storm before runoff can be separated from the baseflow. Frequently, curves of runoff recession are simply sketched with an estimate of duration of runoff obtained from inspection of other storms having the same length of rainfall. However, if the graph shown in Figure 6.5 is constructed, Eq. (6.1) can be used to plot the runoff recession curve AB using the value of K given by the lower line. Curves CD and EF are an extrapolation of the recession curves, but point B must be established by an estimation of time duration of runoff.

THE UNIT HYDROGRAPH

A unit hydrograph is defined as the hydrograph that would result from 1 inch of excess rainfall occurring during a storm of particular duration (4). The unit hydrograph is assumed to be representative of the runoff process for a watershed. Three independent assumptions are made in the concept of the unit hydrograph:

1. For a given drainage basin, the duration of surface runoff is essen-

tially constant for all storms of uniform intensity and the same duration regardless of the total volume of surface runoff.

2. For a given drainage basin, two storms of uniform intensity and the
   same duration would produce rates of surface runoff at time t (after
   the storm's beginning) in direct proportion to the total volumes of
   surface runoff.
3. The time distribution of surface runoff from a given storm period is
   independent of concurrent runoff from previous storm periods.

A storm of uniform intensity is one in which the rainfall occurs at a uniform intensity over the entire drainage basin.

Because of assumption 3 unit hydrograph theory is often referred
to as a linear theory, for which the principle of superposition is applicable. The superposition principle is basic to the construction and use of
the unit hydrograph. In order to obtain a unit hydrograph, we do not
search for a storm in which 1 inch of effective rainfall was accumulated
in the desired storm duration, but instead we make use of the superposition principle at the outset. The steps are as follows (see Fig. 6.7 and
Table 6.2):

1. Search rainfall records to find storms of the desired duration.
2. Locate the associated discharge hydrographs from streamflow records
   (columns 1 and 2, Table 6.2).
3. Separate baseflow from surface runoff for the chosen streamflow hydrographs (columns 3 and 4, Table 6.2).
4. The volume of surface runoff is determined for each hydrograph by
   integration of the runoff portion. The trapezoid rule is used in
   Table 6.2.
5. The average depth of rainfall excess for the storm is determined (1.64
   inches in Table 6.2). The average depth of rainfall excess for the
   unit hydrograph is to be 1 inch. The unit hydrograph is constructed
   by
   a. Maintaining the same time coordinates as the measured hydro-
      graph.
   b. Dividing each of the runoff coordinates by the average depth of
      rainfall excess (1.64 inches in Fig. 6.7) to find the ordinates of the
      unit hydrograph (column 5 of Table 6.2).
6. The result is the unit hydrograph. Unfortunately storms of the same
   duration and intensity may result in different unit hydrographs due
   to different storm orientation, sequence of rainfall increments, and

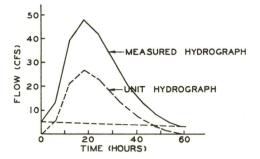

FIG. 6.7. Construction of a unit hydrograph from a measured hydrograph.

TABLE 6.2. Hydrograph resulting from a 3-hour storm on a 685-acre watershed

Time (hrs)	Measured flow (cfs)	Base-flow (cfs)	Runoff portion (cfs)	Unit hydrograph (cfs)
(1)	(2)	(3)	(4)	(5)
0	5.0	5.0	0	0
6	12.9	4.5	8.4	5.1
12	39.4	4.0	35.4	21.6
18	48.1	3.9	44.2	27.0
24	42.3	3.8	38.5	23.5
30	31.5	3.6	27.9	17.0
36	20.8	3.3	17.5	10.7
42	13.3	3.2	10.1	6.2
48	8.3	3.1	5.2	3.2
54	4.8	3.1	1.7	1.0
60	3.0	3.0	0	0
Total			188.9	

Note: Volume of runoff $= \Sigma \ \frac{1}{2} \ (Q_t + Q_{t+\Delta t})\Delta t = \Delta t \Sigma Q_t$
$$= (\tfrac{1}{4} \ \text{day}) \ (188.9 \ \text{cfs}) = 47.225 \ \text{cfs-day}$$
$$= 1123.77 \ \text{acre-inches}$$

Average depth of rainfall excess over 685-acre watershed $= 1123.77$ acre-inches/685 acres $= 1.64$ **inches**

condition of the drainage area. Hence several unit hydrographs should be constructed and the curves averaged. This process is shown in Figure 6.8 in which the results of Figure 6.7 constitute the unit hydrograph from storm 3. Straight numerical averaging of the discharges and times of the unit hydrographs is not appropriate. The time to peak is averaged and the peak discharge is averaged. This average point is plotted, and a hydrograph is sketched that maintains the correct shape and volume.

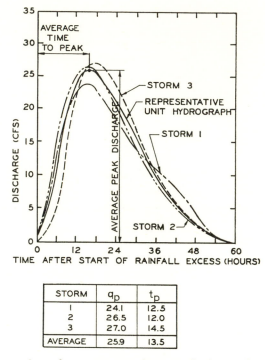

STORM	q_p	t_p
1	24.1	12.5
2	26.5	12.0
3	27.0	14.5
AVERAGE	25.9	13.5

FIG. 6.8. Construction of a representative unit hydrograph. Duration of all storms is approximately three hours.

APPLICATION OF THE UNIT HYDROGRAPH

The unit hydrograph was constructed for a storm of a specified duration and represents the runoff hydrograph that would result if 1 inch of rainfall excess fell in a storm of that duration. The unit hydrograph is used to construct hydrographs for storms of other durations and with other depths of rainfall excess.

The easiest hydrograph to construct is that resulting from a storm with a duration the same as that of the unit hydrograph but in which something other than 1 inch of rainfall excess occurs. In this case the hydrograph is constructed by reversing the process of generating the unit hydrograph; the ordinates are multiplied by the inches of rainfall excess, keeping the time coordinate unchanged.

A more complex procedure is necessary to construct the hydrograph resulting from a storm of duration different from that of the unit hydrograph. This process is depicted in Figure 6.9. The rainfall excess is grouped into increments, each increment having a duration equal to

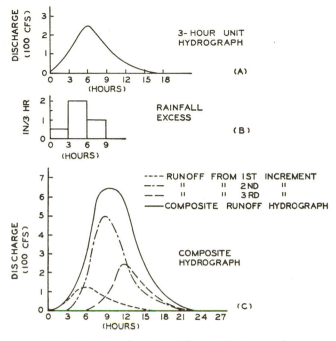

FIG. 6.9.   Application of the unit hydrograph.

the duration of the unit hydrograph as shown in Figure 6.9b. Thus to use a storm of 9-hour duration with a 3-hour unit hydrograph, the first 3 hours of the storm are put in the first increment, the second 3 hours in the second, and the third 3 hours in the third. The unit hydrograph is applied to each of the three increments independently, and a hydrograph is developed representing the flow resulting from each increment of rainfall. These are shown as dotted lines in Figure 6.9c.

The individual hydrographs must be combined to determine the composite rainfall. During the first increment only runoff from that increment occurs. With the start of the second increment of rainfall the composite hydrograph represents flow from both the first and second increments, so the discharges are added. With the start of the third increment of rainfall all three rainfall increments contribute to the flow, so all three discharges are added.

SYNTHETIC UNIT HYDROGRAPHS

The number of streams on which gaging stations are located is very small in comparison to the total of streams and rivers, and existing gaging stations are not always located at the point of interest. As a result we are frequently required to synthesize a unit hydrograph. Although many methods have been proposed for obtaining synthetic hydrographs (5), only two will be presented here.

SNYDER UNIT HYDROGRAPH

Many unit hydrographs were generated and analyzed by Snyder (6) in order to develop a set of empirical relationships for the significant points on a representative unit hydrograph. The most widely used relationships are:

$$t_p = 0.95C_t(LL_{ca})^{0.3} + 0.74t_r \qquad (6.4)$$

$$q_p = 640C_p/t_p \qquad (6.5)$$

$$T = 3 + t_p/8 \qquad (6.6)$$

in which

T = base time in days
t_p = time in hours to the hydrograph peak
t_r = rainfall duration in hours
q_p = peak rate of discharge in cfs/square mile
L_{ca} = river mileage from the point of interest to a point approximately opposite the centroid of the area
L = river mileage from the point of interest to upstream limits of the drainage area
C_t = empirical coefficient accounting for slope in the drainage basin

Figure 6.10 illustrates the variables involved in the Snyder hydrograph. Some values for the coefficients are:

Area	C_p	C_t
Appalachian Highlands	0.63	2.0
Southern California (extremes)	0.94	0.4
Eastern Gulf of Mexico (extremes)	0.61	8.0

Equations (6.4), (6.5), and (6.6) are insufficient to describe the shape of the unit hydrograph. Figure 6.11 provides additional dimensions to be used in shaping the bell portion of the curve (7). The constant of three days in Eq. (6.6) indicates that the Snyder unit graph is not applicable to small drainage basins where the period of runoff is of the order of hours.

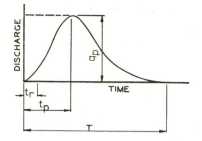

FIG. 6.10.   Snyder unit hydrograph.

TRIANGULAR HYDROGRAPH

In cases where a less precise geometrical shape for the unit hydrograph will be adequate, the triangular unit hydrograph can be
used. The relationships used to generate the triangular hydrograph were
developed by the U.S. Department of Agriculture, Soil Conservation Service (8). This method has the advantage of being extremely convenient.
For a unit hydrograph the volume of runoff is to be

$$V = (1 \text{ inch}) \text{ (area of watershed)}$$

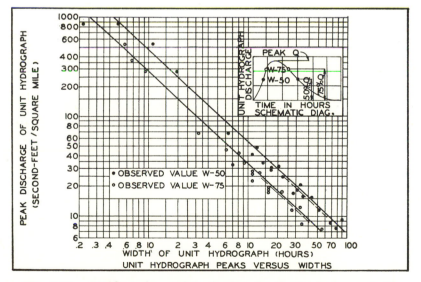

FIG. 6.11.   Widths to be used in shaping the Snyder hydrograph (7).

From the geometry of the triangle the volume is also given by

$$V = (1/2)q_p T \qquad (6.7)$$

so

$$q_p = 2V/T \qquad (6.8)$$

It is assumed that

$$T = (1 + H)(t_p + t_r/2) \qquad (6.9)$$

and

$$t_p = at_c + 0.5t_r \qquad (6.10)$$

where t_c is the time of concentration in hours, determination of which is discussed below. Both a and H are coefficients that depend on watershed characteristics; average values of $H = 1.67$ and $a = 0.6$ are widely used. Figure 6.12 illustrates the triangular hydrograph.

When the triangular unit hydrograph is used to synthesize a streamflow hydrograph from a given storm, the storm should be divided into duration periods t_r such that t_r is less than $0.2t_c$. If t_r is chosen to be larger than $0.2t_c$, the time to peak on the composite hydrograph is not realistic. This effect is illustrated in Figure 6.13.

The time of concentration is the time required for a drop of water to flow from the most distant part of the watershed to the point of interest. This time is best estimated by dividing the channel into reaches, each of which is uniform throughout its length. The velocity of flow and travel time for each reach are estimated, using an appropriate open-channel flow relation, such as Manning's equation, for bank-full flow. The sum of the travel times for the reaches is taken to be the time of

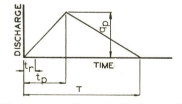

FIG. 6.12. Triangular hydrograph.

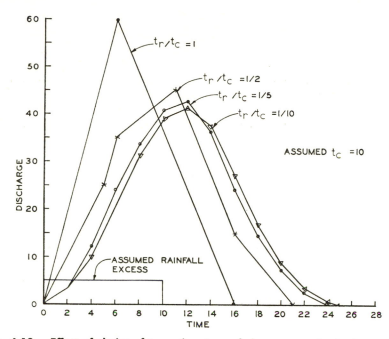

FIG. 6.13. Effect of choice of t_r on time to peak for a composite hydrograph.

concentration. Some representative flow velocities are shown in Table 6.3.

Many empirical equations have been proposed for quick estimates of time of concentration. The most widely used are indicated below.

$$t_c = (11.9L^3/h)^{0.385} \tag{6.11}$$

TABLE 6.3. Representative flow velocities

Slope (%)	Flow velocity (ft/sec)			
	Woodlands (upper watershed)	Pasture (upper watershed)	Natural channel (not well defined)	Natural channel
1–2	1	1.5	1	2
2–4	1.5	2	1	3
4–6	2	3	3	4
6–10	3	4	5	5
10–15	3.5	4.5	8	...

in which

t_c = time of concentration (hours)
L = stream length (miles)
h = difference in elevation in feet between upper and lower limits of the drainage basin

Equation (6.11) is due to Kirpich (9). Hathaway (10) proposed the equation

$$t_c{}^{2.14} = 2Ln/3\sqrt{S} \qquad (6.12)$$

in which

t_c = time of concentration (minutes)
L = channel length (feet)
S = mean slope of the basin
n = Manning's roughness coefficient

Manning's n varies mainly with the channel material and construction. Some representative values are as follows:

Channel	n
Clean, straight banks	0.025–0.033
Same but weeds	0.030–0.040
Winding, some pools	0.33–0.045
Sluggish reaches	0.050–0.080
Very weedy reaches	0.075–0.150

LIMITATIONS OF THE UNIT HYDROGRAPH

At the beginning of the discussion of unit hydrographs three assumptions relative to the concept were set forth. In the subsequent development only storm duration was considered as a variable.

If the storm actually occurs with uniform intensity over the entire drainage basin, as is assumed, its orientation is immaterial. In actuality, a storm virtually never occurs in such a fashion. Storms generally move across a basin, and their direction of movement will play a role in determining the nature of runoff. The larger the drainage basin, the less the chance that any storm will satisfy the condition of uniform intensity. Therefore, unit hydrographs are considered to be applicable only to drainage basins having an area of less than 1000 square miles and preferably smaller if accuracy is quite important. On a large drainage basin substantial variations in soil conditions may also exist.

EQUATIONS FOR DETERMINATION OF PEAK FLOW

Before collection of hydrologic data was begun on the present scale, it was still necessary for engineers to design drainage structures. As a result a multitude of empirical equations for estimating flood flows were formulated and used extensively. Mead (11) lists many of these equations. The most commonly encountered of these formulas include

$$\text{Meyers equation } Q_p = bA^{0.5} \qquad (6.13)$$
$$\text{Talbot equation } a = cA^{0.75} \qquad (6.14)$$
$$\text{Burkli-Ziegler equation } Q_p = Ami(S/A)^{0.5} \qquad \textbf{(6.15)}$$

where

$$A = \text{drainage area}$$
$$a = \text{culvert area}$$
$$S = \text{stream slope}$$
$$i = \text{rainfall intensity}$$
$$Q_p = \text{peak discharge}$$

The constants b and c must account for all the parameters of rainfall-runoff relations, frequencies, and the factors that control hydrograph shape. In the Burkli-Ziegler equation m has the same broad significance except that rainfall intensity and duration are incorporated in i. Because these equations have been used primarily for small drainage areas, they probably cannot be blamed for any spectacular failures. However, their continued use is inadvisable with the more dependable methods available today as a result of several decades of data collection and analysis.

THE RATIONAL METHOD

The so-called rational equation

$$Q_p = CiA \qquad (6.16)$$

has been in use for more than 200 years and in England is known as the Lloyd-Davies equation (12). The definition of the terms is as follows:

Q_p = peak runoff rate (acre-inches/hour)
i = average rainfall intensity (inches/hour)
A = drainage area (acres)
C = runoff coefficient depending upon the characteristics of the drainage area

Since peak flow rate is computed, the rational method assumes that rainfall duration is greater than the time of concentration. It also tacitly

TABLE 6.4. Runoff coefficients for use in the rational equation

Description of area	C
Business	
Downtown	0.70–0.95
Neighborhood	0.50–0.70
Residential	
Single-family	0.30–0.50
Multiunits, detached	0.40–0.60
Multiunits, attached	0.60–0.75
Residential suburban	0.25–0.40
Apartment	0.50–0.70
Industrial	
Light	0.50–0.80
Heavy	0.60–0.90
Parks, cemeteries	0.10–0.25
Playgrounds	0.20–0.35
Railroad yard	0.20–0.35
Unimproved	0.10–0.30
Character of surface	
Pavement	
Asphalt and concrete	0.70–0.95
Brick	0.70–0.85
Roofs	0.75–0.95
Lawns, sandy soil	
Flat, up to 2% grade	0.05–0.10
Average, 2%–7% grade	0.10–0.15
Steep, over 7%	0.15–0.20
Lawns, heavy soil	
Flat, up to 2% grade	0.13–0.17
Average, 2%–7% grade	0.18–0.22
Steep, over 7%	0.25–0.35

assumes that the rate of runoff does not increase after the time of concentration is reached (infiltration is constant). This approximation is probably valid for a small area if the time of concentration is accurately known. In any event, the rational formula is in use by 90% of all state highway departments (13).

Coefficients for use in Eq. (6.16) are given in Table 6.4 (14). The lower portion of the table is used when it is desirable to find an average value for C based on the percentage of different types of surface in the drainage area. These coefficients are applicable for storms of 5- to 10-year return periods. Storms of longer return periods (higher intensity) will require the use of larger coefficients because infiltration and other losses are proportionately less.

Any desired frequency of storm can be constructed and used with the rational equation. In determining the peak value of runoff, the largest value of i (having a duration at least equal to the time of concentration) is used. A curve of average intensity versus duration is constructed for the desired storm frequency as discussed in Chapter 4. The

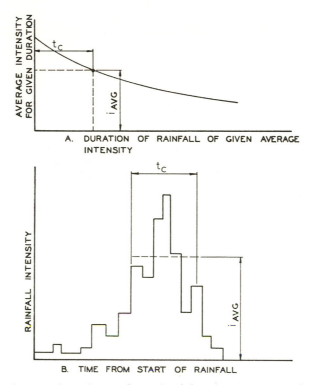

FIG. 6.14. Average intensity as determined from a given storm for use in the rational equation.

average intensity for a duration equal to the time of concentration is read directly from that graph. Figure 6.14 shows the location of this period of average intensity within the storm actually considered as well as an illustration of an average-intensity duration curve. The time of concentration can be determined using one of the procedures discussed earlier in this chapter.

CONCLUSION

Many relatively complex hydrologic problems can be considered, and many of the subleties of rainfall and runoff problems can be explored with the concepts developed thus far. The hydrologic analysis associated with spillway design is discussed in a series of papers by the American Society of Civil Engineers, Task Force on Spillway Design Floods (15, 16, 17, 18). Application of hydrologic analysis to culvert design is detailed by Chow (19), whereas application to design of storm sewers is given in Tholin and Keifler (20) and American Society of Civil Engineers (21). Hydrographs and unit hydrographs and their subtleties are examined by Barnes (22, 23), Dooge (24), and Chow (5).

PROBLEMS

6.1.   The figures below represent streamflow from a 100-acre drainage area. Construct a unit hydrograph for this basin.

Time (hrs)	Flow (cfs)		Time (hrs)	Flow (cfs)
M	20		10	31
2	18		N	26
4	16		2	21
6	26		4	19
8	36		6	17

6.2.   The hydrograph in Figure P6.2 was measured. Construct a unit hydrograph. The area of the watershed is 2,720,000 square feet.

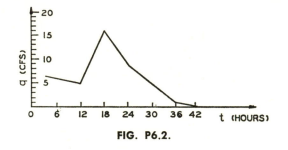

FIG. P6.2.

6.3.   A storm and a unit hydrograph are given in Figure P6.3. Determine the runoff hydrograph.

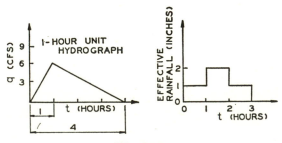

FIG. P6.3.

6.4. Construct a unit hydrograph for a river in your vicinity or for one of the rivers listed in the Appendices. Use actual data on streamflow and precipitation.

6.5. Construct synthetic unit hydrographs, using the Snyder and triangular methods, for a watershed in your vicinity or for a watershed listed in the Appendices. Compare with the results of Problem 6.4.

6.6. Using one of the unit hydrographs of Problem 6.4 or 6.5, construct the runoff hydrograph for the 50-year storm and the probable maximum storm.

6.7. Three hydrographs derived from separate storms are given in the accompanying table. All storms were 4-hour rains. Drainage area of the basin is 10.5 square miles. Construct the three corresponding unit hydrographs and an average unit hydrograph.

Hours	Storm 1	Storm 2	Storm 3
0	20	40	33
1	20	40	33
2	20	40	33
3	130	65	49
4	385	165	91
5	520	398	206
6	410	505	370
7	330	445	443
8	260	345	433
9	200	260	318
10	155	215	250
11	120	175	200
12	90	135	157
13	70	105	125
14	52	80	95
15	40	65	70
16	35	53	51
17	34	46	35
18	33	45	33
19	33	45	33

6.8. Using the unit graph determined in Problem 6.7, determine the peak flow and time to peak resulting from three successive 4-hour storms having rainfall excesses of 1.5, 0.4, and 0.8 inches respectively.

6.9. Construct a Snyder unit hydrograph and a triangular unit hydrograph for a drainage area of 15 square miles in southern Pennsylvania. The upper and lower ends of the drainage basin are at elevations of 467.0 and 401.0 feet respectively. The main stream length is 13 miles. Stream mileage to a point opposite the centroid of the drainage is 6 miles.

6.10. Using a drainage basin assigned by your instructor, look up rainfall records and associated streamflow records. Using these records, construct a unit hydrograph. Construct a Snyder unit hydrograph and a triangular unit hydrograph for the same stream. Compare the three unit graphs.

6.11. Calculate the runoff from a paved asphalt parking lot in Kansas City, Mo., having an area of 10 acres, a length of 1000 feet, and an average slope of 0.002% for
a. A 5-year, 6-hour storm.
b. A 10-year, 12-hour storm.

REFERENCES

1. Pogge, Earnest. Analysis of bank storage. Ph.D. Diss., Dept. Mech. Hydraul., Univ. Iowa, 1967.
2. Chow, Ven Te, ed. Runoff, Sect. 14. *Handbook of Applied Hydrology.* McGraw-Hill, 1964.
3. Barnes, B. S. Discussion on analysis of run-off characteristics by O. H. Meyer. Trans. ASCE, Vol. 105, 1940.
4. Sherman, L. K. Streamflow from rainfall by the unit-graph method. *Eng. News Rec.,* Vol. 108, Apr. 7, 1932.
5. Chow, Ven Te, ed. *Handbook of Applied Hydrology.* McGraw-Hill, 1964.
6. Snyder, F. M. Synthetic unit graphs. Trans. Am. Geophys. Union, Vol. 19, 1938.
7. U.S. Army, Office of the Chief of Engineers. Engineering manual for civil works. Part 2, Chap. 5, Apr. 1946.
8. U.S. Soil Conservation Service. Hydrology, Chap. 6. *In* National Engineering Handbook, Sect. 4, Part 1, Washington, D.C., 1964.
9. Kirpich, P. Z. Time of concentration of small agricultural watersheds. *Civil Engineering,* Vol. 10, No. 6, June 1940.
10. Hathaway, G. A. Design of drainage facilities. Trans. ASCE, Vol. 110, 1945.
11. Mead, D. W. *Hydrology.* McGraw-Hill, 1950.
12. Chow, Ven Te. Hydrologic determination for waterway areas for the design of drainage structures in small drainage basins. Univ. Illinois Bull., Vol. 59, No. 65, Mar. 1962.
13. Wycoff, R. L.; and Harbaugh, T. E. A survey of the hydrologic design practices of state highway departments. Hydrol. Ser. Bull., Univ. Mo., Rolla, June 1970.
14. American Society of Civil Engineers. Design and construction of sanitary and storm sewers. Manuals and Reports of Engineering Practice No. 37, 1970.
15. Banks, H. O. Hydrology of spillway design: Introduction. *ASCE J. Hydraul. Div.,* Vol. 90, Hy 3, May 1964.
16. Snyder, F. F. Hydrology of spillway design: Large structures—Adequate data. *ASCE J. Hydraul. Div.,* Vol. 90, Hy 3, May 1964.
17. Koelzer, V. A.; and Bitoun, M. Hydrology of spillway design floods: Large structures—Limited data. *ASCE J. Hydraul. Div.,* Vol. 90, Hy 3, May 1964.

18. Ogrosky, H. O. Hydrology of spillway design: Small structures—Limited data. *ASCE J. Hydraul. Div.*, Vol. 90, Hy 3, May 1964.

19. Chow, Ven Te. Hydrologic design of culverts. *ASCE J. Hydraul. Div.*, Vol. 88, Hy 2, Mar. 1962.

20. Tholin, A. L.; and Keifer, C. J. The hydrology of urban runoff. Trans. ASCE, Vol. 125, 1960.

21. American Society of Civil Engineers. Design and construction of sanitary and storm sewers. Manuals and Reports on Engineering Practice No. 37, 1969.

22. Barnes, B. S. Unit-graph procedures. USDI, Bureau of Reclamation, Washington, D.C., 1952.

23. ———. Consistency in unit graphs. *ASCE J. Hydraul. Div.*, Vol. 85, Hy 8, Aug. 1959.

24. Dooge, J. C. I. A general theory of the unit hydrograph. *J. Geophys. Res.*, Vol. 64, No. 2, Feb. 1959.

CHAPTER SEVEN

FLOOD ROUTING

FLOOD ROUTING is the analytical process of determining the shape of a flood hydrograph at a particular location in a channel, reservoir, or lake resulting from a measured or hypothesized flood at some other location. Such calculations are necessary in establishing the height of a flood peak at a downstream location; estimating the protection that results from construction of a reservoir; and determining required levee heights for flood protection, adequacy of spillways, and any other flood-related calculations.

The basis of the flood routing procedure can be seen by considering the region between the upstream and downstream points as a *black box* (Fig. 7.1). Thus the black box replaces a reach (length) of channel or a reservoir. The inflow hydrograph *I* flowing into the box is known either as a measured or as a synthesized flood (an example is given in column 2 of Table 7.2). The outflow from the black box is to be determined. The principle of continuity of fluid flow is used to provide the basic equation for solution of the problem. It may be stated as:

Inflow volume in time increment Δt
— outflow volume in time increment Δt
= change in volume of water stored inside the black box (7.1)

There will obviously be a change in the amount of water inside the

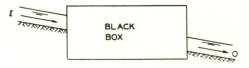

FIG. 7.1.   Simplified stream model.

black box because the depth of flow will increase as the flood passes. The continuity equation can be expressed as:

$$I\Delta t - O\Delta t = \Delta S \qquad (7.2)$$

where

$$I \text{ and } O = \text{the rates of inflow and outflow respectively}$$
$$S = \text{storage volume}$$

Thus ΔS represents a change in storage volume occurring during the time interval Δt. The values of I are known and tabulated for various times; therefore, Δt is implied by the time interval of the tabulation. The outflow O is to be determined, but S is not yet established as a function of time. There are two unknowns in the equation, S and O. The storage is a characteristic of the black box. The problems of establishing the storage differ, depending on whether the box consists of a length of river channel, a dam and associated reservoir, or a special combination of both. The case of the reservoir will be considered first.

RESERVOIR ROUTING

For a reservoir the water surface is assumed to be level at all times, although that may not always be the case. The following are known initially:

1. the inflow hydrograph
2. the depth of water (or elevation of the water surface) in the reservoir before the flood arrives $(t = 0)$
3. the outflow from the reservoir before the flood arrives $(O, t = 0)$

In addition to these known quantities a substantial amount of information is available because the reservoir is a man-made structure. The volume of storage as a function of water surface elevation can be established from a topographic map if the water surface is assumed to be always level. An example of such information is shown in Table 7.1 (1). Figure 7.2 illustrates the reservoir and spillway. Also, certain information describing the manner in which water can be released from the reservoir is available, i.e., through the turbines or outlet works or over the spillway. These releases are controlled in part by an operator. The uncontrolled releases, those going freely over the spillway, are determinate and depend upon the depth of flow over the spillway and the

TABLE 7.1. Values of storage, outflow, and $(2S/\Delta t) + O$ for $\Delta t = 15$ minutes

Stage (h) (ft)	Storage (S) (cu ft)	Outflow (O) (cfs)	$2S/\Delta t + O$ (cfs)
	(1)	(2)	(3)
10.29	530	0	1.18
10.3	535	0.01	1.20
10.4	680	0.32	1.83
10.6	1,050	2.55	4.88
10.8	1,550	7.50	10.9
11.0	2,240	14.4	19.4
11.2	3,210	21.6	28.7
11.4	4,580	28.9	39.1
11.6	6,430	36.3	50.6
11.8	8,800	44.3	63.9
12.0	11,900	53.0	79.4
12.2	16,800	62.0	99.3
12.4	23,400	71.5	123
12.6	31,600	81.5	152
12.8	41,800	92.0	185
13.0	53,600	103	222
13.2	67,900	114	265
13.4	86,200	126	318
13.6	111,000	138	385
13.8	143,000	150	468
14.0	190,000	163	585
14.2	256,000	177	746
14.4	352,000	191	973
14.6	493,000	205	1,300
14.8	658,000	219	1,680
15.0	952,000	233	2,350

Source: Carter and Godfrey (1).
Note: Column 3 is computed directly from the values for S and O as given in columns 1 and 2 for the same value of h.

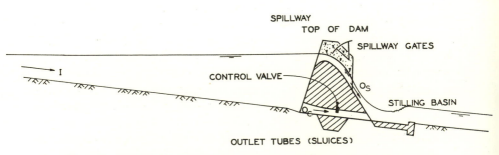

FIG. 7.2. Reservoir and associated dam, spillway, and outlet tubes.

spillway geometry. The depth of flow over the spillway will depend in turn upon the depth of water in the reservoir. An example of a spillway rating curve (the relationship between discharge and water surface elevation or stage) is given in Table 7.1.

As a result of the foregoing discussion the following expressions can be written:

$$S = S(h)$$
$$O = O_c + O_s(h) \qquad\qquad (7.3)$$

in which

$$h = \text{stage}$$
$$O_c = \text{controlled discharge}$$
$$O_s(h) = \text{uncontrolled discharge}$$

In principle one could write expressions for $S(h)$ and $O_s(h)$, substitute these into the continuity equation written in differential form, and integrate the result. The solution would give reservoir elevation h as a function of time. Actually the resulting differential equation is usually impossible to solve exactly and one must resort to numerical calculations. In order to perform the numerical calculations, let

$$I = (I_t + I_{t+\Delta t})/2$$
$$O = (O_t + O_{t+\Delta t})/2 = (O_{c_t} + O_{c_{t+\Delta t}} + O_{s_t} + O_{s_{t+\Delta t}})/2 \qquad (7.4)$$
$$\Delta S = S_{t+\Delta t} - S_t$$

Note that I and O now represent average inflow and outflow rates occurring during time interval Δt. The quantity ΔS represents the change in volume of water stored occurring during the time interval Δt. If the reservoir surface were rising, ΔS would be positive in sign.

Take $t = 0$ as the initial time; I and the O_c are known for all time, but O_s and S are known only at $t = 0$. The unknowns are $O_{s_{t+\Delta t}}$ and $S_{t+\Delta t}$. Substitution of Eq. (7.4) into Eq. (7.2) (the continuity equation), placing all of the known quantities on the left-hand side of the equation and collecting terms, results in

$$(I_t + I_{t+\Delta t}) - (O_{c_t} + O_{c_{t+\Delta t}}) + \left(\frac{2S_t}{\Delta t} - O_{s_t}\right) = \frac{2S_{t+\Delta t}}{\Delta t} + O_{s_{t+\Delta t}} \qquad (7.5)$$

Setting $t = 0$ and substituting the known quantities into the left side of Eq. (7.5) yields a numerical quantity for the right-hand side. This still does not yield a value for $O_{s_{t+\Delta t}}$. Both the storage and the discharge, the variables on the right side of Eq. (7.5), are functions of h, however, and a table of

$$(2S/\Delta t) + O_s = f(h) \tag{7.6}$$

can be constructed, as in column 3 of Table 7.1. Thus solutions of Eq. (7.5) result in the value of $(2S/\Delta t) + O_s$ at time $t + \Delta t$. This value corresponds to a particular value of h, the elevation of the reservoir surface at time $t + \Delta t$. That elevation h is determined by interpolation in column 3 of Table 7.1. When h at $t + \Delta t$ is obtained, the storage and outflow at time $t + \Delta t$ can be read from Table 7.1.

TABLE 7.2. Typical calculations for routing a flood through a reservoir

Time	I = inflow (cfs)	$I_t + I_{t+\Delta t}$	O = outflow (cfs)	$2S/\Delta t - O$	$2S/\Delta t + O$	h = stage (ft)
(1)	(2)	(3)	(4)	(5)	(6)	(7)
0:00	0	...	0	1.18	1.18	10.29
:15	1.44	1.44	0.84	0.94	2.62	10.47
:30	4.34	5.78	4.04	−1.36	6.72	10.67
:45	8.90	13.24	8.30	−4.72	11.88	10.82
1:00	18.7	27.6	17.2	−11.5	22.9	11.08
1:15	38.8	57.5	33.5	−21.0	46.0	11.52
1:30	106	144.8	71.6	−19.4	123.8	12.40
1:45	216	322	123	57	303	13.35
2:00	291	507	161	242	564	13.97
2:15	320	611	185	483	853	14.31
2:30	325	645	198	732	1,128	14.50
2:45	309	634	208	950	1,366	14.64
3:00	285	594	214	1,116	1,544	14.73
3:15	260	545	219	1,223	1,661	14.80
3:30	235	495	220	1,278	1,718	14.81
3:45	211	446	220	1,284	1,724	14.81
4:00	188	399	219	1,245	1,683	14.80
4:15	165	353	217	1,164	1,598	14.77
4:30	145	310	212	1,050	1,474	14.70
4:45	129	274	206	912	1,324	14.61
5:00	116	246	199	760	1,158	14.51
5:15	106	222	191	600	982	14.40
5:30	96.4	202.4	181	440.4	802.4	14.26
5:45	88.0	184.4	166	292.8	624.8	14.04
6:00	80.2	168.2	149	163.0	461.0	13.78
15:00	0.75	1.82	0.83	0.95	2.61	10.47
15:15	0.47	1.22	0.54	1.09	2.17	10.43
15:30	0.22	0.69	0.30	1.18	1.78	10.39
15:45	0	0.22	0.11	1.18	1.40	10.33
16:00	0	0	0	...	1.18	10.29
Total	4,504		4,504			

Source: Carter and Godfrey (1).

This process was carried out in Table 7.2 (1) for $t = 0$ to get conditions at $t = 0 + \Delta t$. The step to $t = 0 + 2\Delta t$ can now be made. Note that the quantity $(2S_t/\Delta t) - O_{s_t}$ will be needed and that these quantities again depend only on h. Another column can be constructed to give this parameter as a function of h, or it can be computed from

$$[(2S/\Delta t) + O_s] - 2O_s = 2S/\Delta t - O_s \qquad (7.7)$$

As the solution is constructed step by step, it will become obvious that much interpolation is required; for ease in doing this the values in Table 7.1 are frequently graphed. Table 7.2 shows a handy form for this step-by-step solution process. In addition to the inflow hydrograph the underlined values are known at the outset. Using the known quantities,

$$(2S_{t+\Delta t}/\Delta t) + O_{s_{t+\Delta t}} \qquad (7.8)$$

is computed. The $t + \Delta t$ row is filled by using Table 7.1, and the process is repeated for another time step Δt.

Details of the construction of Table 7.2 follow:

Column	Explanation
1	Specified time interval (independent variable).
2	Inflow hydrograph known at the outset.
3	Sum of two adjoining flows in column 2, written opposite the end of the time interval considered.
4	For this example, outflow is a function of h only. The value of O in column 4 is determined by the value h in column 7 and is obtained from Table 7.1. At $t = 0$ we are given that $h = 10.29$ and thus $O = 0$.
5	Obtained by subtracting $2O$ from column 3 of Table 7.1. Corresponds to the beginning of the time interval. At time 0:15, $h = 10.47$ which corresponds to $O = 0.84$ and $(2S/\Delta t) + O = 2.62$; thus $(2S/\Delta t - O = 2.62 - 2(0.84) = 0.94$.
6	Obtained by adding the value on the same line of column 3 to the value on the previous line of column 5. At time 0.30, $5.78 + 0.94 = 6.72$.
7	Obtained by interpolating for h in Table 7.1 using the value of $(2S/\Delta t) + O$ on the same line of column 6. This is the stage at the end of one time step and the beginning of the next.

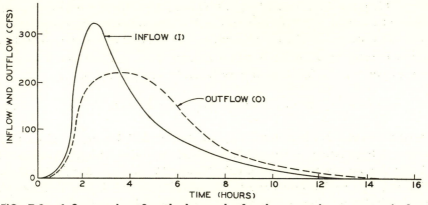

FIG. 7.3.   Inflow and outflow hydrographs for the example on reservoir flood routing (values from Table 7.2).

Figure 7.3 shows the inflow hydrograph (column 2 of Table 7.2) for the previous example as well as the determined outflow hydrograph (column 4 of Table 7.2). This procedure of routing a flood through a reservoir is used to determine the spillway size required to pass a design flood and to determine how much a given reservoir will alleviate a flood. Note that the peak outflow from the reservoir is 220 cfs whereas the peak flow of the inflow hydrograph was 325 cfs. Since the reservoir stage starts and finishes at elevation 10.29, no net storage is accumulated and saved during the flood. Thus the area under the outflow and inflow hydrographs in Figure 7.3 must be equal since they represent equal volumes. A reservoir designed to aid in flood control is operated basically like that in the example. The reservoir stage is purposely held below a specific elevation in order to provide storage for a possible flood. As soon as the flood is passed, the reservoir water surface is lowered as rapidly as possible (without flooding downstream) in order to provide storage for the next flood.

STREAMFLOW ROUTING

Routing a flood down a river channel presents difficulties not encountered in reservoir routing. The basic equation, Eq. (7.2), remains the same. In differential form it is

$$I - O = \frac{dS}{dt} \tag{7.9}$$

For the reservoir the outflow and the storage were both a function of

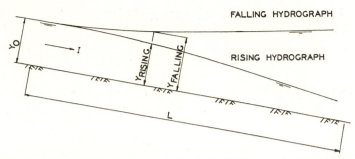

FALLING HYDROGRAPH

RISING HYDROGRAPH

FIG. 7.4. Free surface in an open channel for falling and rising hydrographs.

the depth of water behind the dam. Thus in effect the continuity equation was a relationship between the known inflow and the unknown depth h. In a channel the water surface is not always parallel to the bottom. A long narrow reservoir may thus violate our assumption of a level pool in reservoir routing. When streamflow is increasing, the average free surface slopes more rapidly than the channel bottom, while the opposite occurs for a decreasing streamflow. Figure 7.4 illustrates this phenomenon. Thus the relation between outflow and storage for a particular reach is not obvious and will be different depending upon whether the hydrograph is rising or falling.

One must have some relationship, however, so it is appropriate to assume that the outflow and storage are related in the form of an infinite series (a rather typical thing to do when a simpler or more direct relationship is not obvious).

$$O = aS^n - \sum_{m=1}^{\infty} \left(X^m \frac{d^m S}{dt^m} \right) \qquad (7.10)$$

where a, X, and n are constants which are not known at the outset. As is common in numerical procedures, this series is truncated, discarding all but the first two terms to give

$$O = aS^n - X \frac{dS}{dt} \qquad (7.11)$$

In so truncating, terms of order greater than dS/dt are assumed to be small. This equation can be solved for dS/dt and substituted into the continuity equation, Eq. (7.9), to give

$$S = (1/a)^{1/n} [XI + O(1 - X)]^{1/n} \qquad (7.12)$$

Standard notation is achieved by letting $(1/a)^{1/n} = k$. The exponent n is normally taken to be unity. Thus

$$S = k[XI + (1 - X)O] \qquad (7.13)$$

is the storage equation for the Muskingum method of flood routing (2).

Substituting Eq. (7.13) into Eq. (7.5) (assuming $O_c = 0$) results in

$$O_{t+\Delta t} = C_0 I_{t+\Delta t} + C_1 I_t + C_2 O_t \qquad (7.14)$$

in which

$$C_0 = \frac{0.5\Delta t - kX}{k(1 - X) + 0.5\Delta t}$$

$$C_1 = \frac{0.5\Delta t + kX}{k(1 - X) + 0.5\Delta t} \qquad (7.15)$$

$$C_2 = \frac{k(1 - X) - 0.5\Delta t}{k(1 - X) + 0.5\Delta t}$$

Equation (7.14) is known as the Muskingum routing equation (2).

The values of k and X must be established before the routing equation can be used. This is best done by applying the inverse of the routing process to actual measured hydrographs. Equations (7.5) and (7.13) can be solved to yield

$$k = \frac{\text{storage (numerator } N)}{\text{weighted inflow and outflow (denominator } D)}$$

$$= \frac{0.5\Delta t[(I_t + I_{t+\Delta t}) - (O_t + O_{t+\Delta t})]}{X(I_{t+\Delta t} - I_t) + (1 - X)(O_{t+\Delta t} - O_t)} \qquad (7.16)$$

Successive values of the numerator can be accumulated by using measured flood hydrographs. The denominator can be accumulated from the hydrographs by using various values of X. An example calculation is shown in Table 7.3 and results are plotted in Figure 7.5 (3). The value of X that gives the curve that is most nearly a straight line is assumed to be the appropriate value to use. The nearness to which a straight line is approximated may be taken as a measure of the appropriateness of the approximations used in developing the storage relation,

TABLE 7.3. Determination of coefficients k and x (see Eq. 7.16)

(1) Time Δt = 0.5 day		(2) $I =$ inflow*	(3) $O =$ outflow†	(4) $I_{t+\Delta t} + I_t$	(5) $I_{t+\Delta t} - I_t$	(6) $O_{t+\Delta t} + O_t$	(7) $O_{t+\Delta t} - O_t$	(8) N‡	(9) ΣN	Values of D and ΣD for assumed values of X							
										X = 0		X = 0.1		X = 0.2		X = 0.3	
										(10) D§	(11) ΣD	(12) D§	(13) ΣD	(14) D§	(15) ΣD	(16) D§	(17) ΣD
Feb. 26	AM	2.2	2.0	16.7	9.0	12.3	5.0	1.9	...	5.0	...	5.7	...	6.5	...	7.2	...
	PM	14.5	7.0	42.9	18.7	13.9	4.7	6.1	1.9	4.7	5.0	5.6	5.7	6.5	6.5	7.5	7.2
Feb. 27	AM	28.4	11.7	60.2	28.2	3.4	4.8	8.0	8.0	4.8	9.7	4.6	11.3	4.5	13.0	4.3	14.7
	PM	31.8	16.5	61.5	40.5	-2.1	7.5	5.2	16.0	7.5	14.5	6.7	15.9	5.6	17.5	4.6	19.0
Feb. 28	AM	29.7	24.0	55.0	53.1	-4.4	5.1	0.5	21.2	5.1	22.0	4.1	22.6	3.2	23.1	2.3	23.6
	PM	25.3	29.1	45.7	57.5	-4.9	-0.7	-2.9	21.7	-0.7	27.1	-1.1	26.7	-1.5	26.3	-2.0	25.9
Mar. 1	AM	20.4	28.4	36.7	52.2	-4.1	-4.6	-3.9	18.8	-4.6	26.4	-4.6	25.6	-4.5	24.8	-4.4	23.9
	PM	16.3	23.8	28.9	43.2	-3.7	-4.4	-3.6	14.9	-4.4	21.8	-4.3	21.0	-4.3	20.3	-4.2	19.5
Mar. 2	AM	12.6	19.4	21.9	34.7	-3.3	-4.1	-3.2	11.3	-4.1	17.4	-4.0	16.7	-3.9	16.0	-3.9	15.3
	PM	9.3	15.3	16.0	26.5	-2.6	-4.1	-2.6	8.1	-4.1	13.3	-4.0	12.7	-3.8	12.1	-3.6	11.4
Mar. 3	AM	6.7	11.2	11.7	19.4	-1.7	-3.0	-1.9	5.5	-3.0	9.2	-2.8	8.7	-2.8	8.3	-2.6	7.8
	PM	5.0	8.2	9.1	14.6	-0.9	-1.8	-1.4	3.6	-1.8	6.2	-1.7	5.9	-1.6	5.5	-1.6	5.2
Mar. 4	AM	4.1	6.4	7.7	11.6	-0.5	-1.2	-1.0	2.2	-1.2	4.4	-1.2	4.2	-1.1	3.9	-0.9	3.6
	PM	3.6	5.2	6.0	9.8	-1.2	-0.6	-1.0	1.2	-0.6	3.2	-0.6	3.0	-0.7	2.8	-0.8	2.7
Mar. 5	AM	2.4	4.6	...	...	...	...	...	0.2	...	2.6	...	2.4	...	2.1	...	1.9

Source: U.S. Army, Corps of Engineers (3).
* Inflow is hydrograph at Newcomerstown.
† Outflow has been adjusted to equal the volume of the outflow.
‡ Numerator N is (½) Δt (column 4 − column 5).
§ Denominator D is [column 7 + X (column 6 − column 7)].

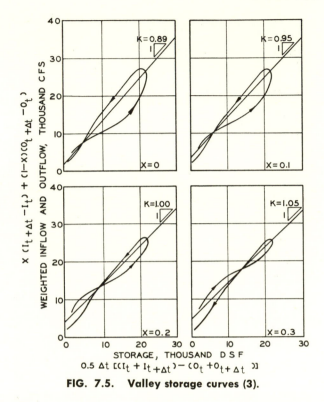

FIG. 7.5.   Valley storage curves (3).

Eq. (7.13). Remember that we initially rejected all but linear terms in forming Eq. (7.11).

The value of k comes automatically from the above process, but values of X must be assumed in order to draw the graphs. The values of k and X can be estimated and their influence on the routing process explored. The constant k has the dimension of time and is equivalent to the time required for an elemental discharge wave to traverse the reach. As a result, the time increment Δt used in the routing process should be approximately equal to k. The value of k can be estimated if the velocity of a discharge wave is known. Use of Seddon's principle with Manning's equation gives the results shown in Table 7.4 where V

TABLE 7.4.   Ratio of flood wave velocity V_w to channel flow velocity V

Channel	V_w/V
Wide rectangular	1.67
Wide parabolic	1.44
Triangular	1.33

Source: Linsley, et al. (2).

is average flow velocity and V_w is the velocity of the discharge wave (2). In Figure 7.4 the difference in water surface slope for a falling and rising hydrograph is actually due in part to the direction of movement of this discharge wave. For a rising hydrograph the wave moves downstream and acts conversely for a falling hydrograph. The velocity V may be obtained from the discharge through a representative cross section of the channel.

The influence of X on the routed hydrograph is shown in Figure 7.6 (3). A value of 1/2 for X results in a hydrograph translated through the reach without change of shape, whereas $X = 0$ results in a reservoir type of storage routing. X may be considered a dimensionless index of the wedge storage in the reach. A unit width of channel is indicated in Figure 7.7 (3). The prism storage is given by Ly_0, where $y =$ depth of flow, and is equal to $O\Delta t$ or, with Δt equal to k, equal to Ok. The wedge storage is given by $L\Delta y/2$, which is proportional to $(I - O)\Delta t$ or $(I - O)k$. The proportionality constant is the quantity X. Thus

$$L\Delta y/2 = X(I - O)k \tag{7.17}$$

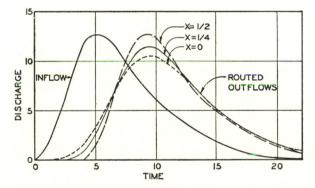

FIG. 7.6.   Routings through four subreaches with $\Delta t = K = 1$ in each subreach

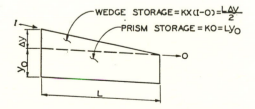

FIG. 7.7.   Portion of length L of a stream channel.

But from the prism storage

$$K = Ly_0/O \qquad\qquad (7.18)$$

So

$$X = O\Delta y/[2(I - O)y_0] \qquad\qquad (7.19)$$

The quantity $(I - O)/\Delta y$ may be found by differentiating Manning's equation. For a wide rectangular channel with small changes in discharge, $X = 0.3$. For triangular channels, X increases uniformly from 0.375 at $\Delta y/y_0 = 0$ to 0.438 at $\Delta y/y_0 = 0.5$.

As an example of the use of the Muskingum method, consider the inflow hydrograph of Table 7.3 and use $k = 1.05$ days and $X = 0.3$ from Figure 7.5. The coefficients (using $\Delta t = 1$ day) of Eq. (7.15) are

$$C_0 = 0.15, \qquad C_1 = 0.66, \qquad C_2 = 0.19$$

The routing is accomplished in Table 7.5. The inflow hydrograph, column 2 (underlined), is known initially. Thus columns 3 and 4 can be computed for all times. The outflow for the first day is known, making it possible to compute the first value in column 5. The routing equation, Eq. (7.14), can be solved to yield the outflow for the second day, and the value of C_2O can be determined for that day. Columns 5 and 6 are determined day by day to yield the outflow hydrograph. Column 6 is the outflow hydrograph from Table 7.3 with which the computed hydrograph should be compared. The difference in the computed peak discharges is about 4% for this particular example.

TABLE 7.5. Example of Muskingum method of flood routing

| | | \multicolumn{5}{c}{$(O_{t+\Delta t} = C_0 I_{t+\Delta t} + C_1 I_t + C_2 O_t)$} | | |
Time (days)	Inflow (cfs)	C_0I ($C_0 = 0.15$)	C_1I ($C_1 = 0.66$)	C_2O ($C_2 = 0.19$)	Computed outflow	Measured outflow
(1)	(2)	(3)	(4)	(5)	(6)	(7)
1	2.2	0.33	1.45	0.38	2.00	2.0
2	28.4	4.26	18.70	1.16	6.09	11.7
3	29.7	4.45	19.60	4.62	24.31	24.0
4	20.4	3.06	13.50	5.19	27.28	28.4
5	12.6	1.89	8.54	3.90	20.58	19.4
6	6.7	1.01	4.42	2.58	13.54	11.2
7	4.1	0.62	2.71	1.45	7.62	6.4
8	2.4	0.36	1.59	...	4.52	4.6

Routings involving tributary inflow can be handled in addition to routings with variable k and X. Details of these processes are given in Carter and Godfrey (1) and U.S. Army (3).

BASEFLOW

In the routing examples used here no mention was made of baseflow. For most flood control reservoirs the baseflow of the stream is an insignificant percentage of the flood hydrograph. However, it should be kept in mind that the baseflow must always be considered; and if it is significant for the stream of interest, the flood hydrograph should be placed on top of a reasonable baseflow curve.

PROBLEMS

7.1. Determine the maximum depth to which water will rise in a hypothetical reservoir where

$$S \text{ (storage in acre-feet)} = 1.0D^2$$
$$O \text{ (outflow in cfs)} = 200D^{1/2}$$
$$D = \text{reservoir depth in feet}$$

and the inflow hydrograph is:

Time	Inflow (acre-in./hr)	Time	Inflow (acre-in./hr)
12:00 P.M.	100	7:00	1200
1:00	100	8:00	1000
2:00	200	9:00	800
3:00	400	10:00	600
4:00	800	11:00	200
5:00	1400	12:00 A.M.	100
6:00	1400		

The depth D at 12:00 P.M. is 0.25 feet.

7.2. A particular reservoir can be considered to be a rectangular prism and thus the storage can be expressed as

$$S = 150h \text{ (cfs-days)}$$

where h is in feet. The outflow is controlled by a weir which discharges according to the relation

$$O = 50h \text{ (cfs)}$$

a. Set up an algebraic expression for $(2S/\Delta t) - O$ as a function of h, using $\Delta t = 2$ days.

b. Repeat part (a) to get an expression for h as a function $(2S/\Delta t) + O$.

c. Route the inflow hydrograph below, finding the head h on the weir as a function of time. The initial elevation h is 1 foot.

Time (days)	I (cfs)
0	50
2	300
4	500
6	200
8	40

7.3. Using the following record of inflow and outflow from a reservoir observed during a particular flood (Fig. P7.3), plot a graph of water surface elevation versus time. The reservoir water surface is at an elevation of 40 feet when inflow starts.

Time (hrs)	Inflow (cfs)	Outflow (cfs)
0	120	0
4	240	24
8	480	96
12	360	240
16	120	240
20	120	240
24	120	240

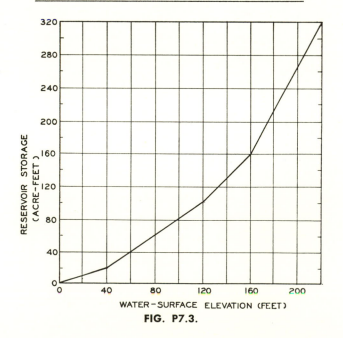

FIG. P7.3.

7.4. A dam and associated reservoir are being planned for flood protection purposes slightly upstream from a large suburban area. The runoff hydrograph for the 24-hour maximum probable storm on Indian Creek is shown along with the spillway rating curve (Fig. P7.4A), the controlled outflow rating curve (Fig. P7.4B), and the reservoir capacity and surface area curves (Fig. P7.4C). Route the maximum probable flood through the reservoir to determine if the spillway is adequate.

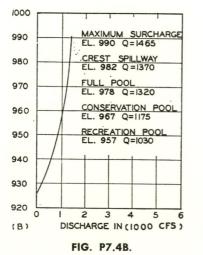

(A)   DISCHARGE (1000 CFS)

FIG. P7.4A.

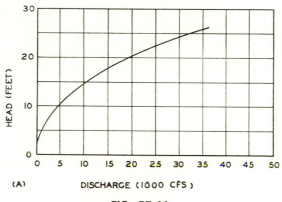

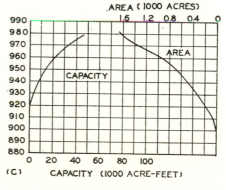

(B)   DISCHARGE IN (1000 CFS)   (C)   CAPACITY (1000 ACRE-FEET)

FIG. P7.4B.   **FIG. P7.4C.**

REFERENCES

1. Carter, R. W.; and Godfrey, R. G. Storage and flood routing. U.S. Geological Survey Water Supply Paper 1543-B, Washington, D.C., 1960.
2. Linsley, R. K.; Kohler, M. A.; and Paulhus, J. L. H. *Hydrology for Engineers.* McGraw-Hill, 1958.
3. U.S. Army, Corps of Engineers. Routing of floods through river channels. EM 1110-2-1408, Mar. 1960.

CHAPTER EIGHT
RESERVOIR YIELD

MOST OF THE EMPHASIS of the preceding chapters has been on peak discharges. Many hydraulic structures are built to store water for later use. The determination of the capabilities of reservoirs to provide water is termed analysis of reservoir yield. In this chapter three general means of determining reservoir yield will be discussed. The first is a simulation approach, whereas the second two are probabilistic in nature.

SIMULATION METHODS

The simplest simulation methods are based on the unrealistic assumption that the flows of the past will be repeated identically in the future. One may simulate the operation of the reservoir by analytically routing the past record of flow into the reservoir, extracting withdrawal and losses and keeping track of the quantity of water in storage. That is, the continuity equation

$$\text{Inflow volume} - \text{outflow volume} = \text{change in storage volume}$$

is solved numerically for specific periods of time.

As an example, consider a reservoir on the Little Blue River having a maximum storage of 200,000 acre-feet (which is equivalent to 277 cfs-years). Losses due to seepage and evaporation will be neglected in order to simplify the example. The mean annual flows are listed in column 2 of Table 8.1. Assume that the reservoir is empty at the end of 1948. If the total inflow is assumed to occur in one increment during the spring, the reservoir would contain 168 cfs-years (cfs-years is a handy though unusual unit of volume) of water in storage at the end of the spring of 1949. From this storage the demand plus losses must be met. If the de-

TABLE 8.1. Reservoir simulation, Little Blue River

Year	Inflow volume (cfs-yrs)	Year-end storage (cfs-yrs)	Spill (cfs-yrs)	Demand not met (cfs-yrs)
(1)	(2)	(3)	(4)	(5)
1948		0		
1949	168.0	93.0		
1950	101.0	119.0		
1951	176.0	202.0	18	
1952	127.0	202.0	52	
1953	40.2	167.2		
1954	26.1	118.3		
1955	59.5	102.8		
1956	11.5	39.3		
1957	22.8	0		12.9
1958	131.0	56.0		
1959	45.6	26.6		
1960	83.3	34.9		
1961	279.0	202.0	36.9	
1962	202.0	202.0	127.0	
1963	43.7	170.7		
1964	70.1	165.8		
1965	162.0	202.0	50.8	

Note: Maximum storage available = 200,000 acre-ft (277 cfs-yrs). Annual demand = 75 cfs-yrs.

mand is 75 cfs-years, th : year-end storage would be 93 cfs-years. To the 93 cfs-years will be added 101 cfs-years in the spring of 1950. The demand will leave a year-end storage of 119 cfs-years. The inflow in the spring of 1951 results in a midyear storage of 295 cfs-years. The reservoir, however, will only hold 277 cfs-years; therefore, 18 cfs-years must be spilled and the demand will be taken from the full reservoir, leaving a storage of 202 cfs-years at the end of 1951.

Operation will progress as described until 1957. Storage at the end of 1956 was only 39.3 cfs-years. To this storage was added 22.8 cfs-years in the spring. The total storage from which to meet the demand is only 62.1 cfs-years. Thus the demand could not be completely satisfied, and the reservoir ended the year empty.

Table 8.1 and the preceding discussion gives a simplified description of a simulation analysis of reservoir operation. In most cases it is not possible to consider the flow as entering in the spring and going out in the summer and fall. Demand rates vary from day to day and month to month as irrigation, power, and other water requirements vary. Hence, shorter time intervals must frequently be used (months, weeks, or days, depending on the specific problem).

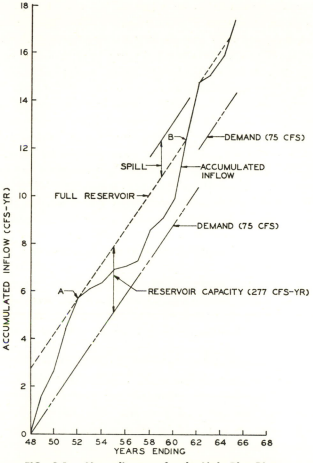

FIG. 8.1. Mass diagram for the Little Blue River.

Losses were not considered in depth for this problem, but should be in most cases. Evaporation and seepage both depend on the volume in storage. Usually, the average of the storage at the beginning and ending of the time period is used to compute the losses.

The procedure indicated in Table 8.1 can also be carried out graphically. Certain modifications are usually made, however. The inflow is assumed to enter at a uniform rate throughout the time period. Similarly, the demand is assumed to occur at a constant rate.

In order to carry out the graphic procedure, a graph of accumulated inflow (cfs-years) is plotted against time (years) as shown in Figure 8.1. The ordinate of this graph is volume and the abscissa is time; hence the slope of a line on this graph represents discharge.

Consider the reservoir to be empty at the end of 1948 and draw a line with a slope equivalent to the demand, 75 cfs. The top of the reservoir will be indicated by a line parallel to the demand line but 277 cfs-years above it. Note that between 1948 and 1952 the slope of the inflow line exceeds that of the demand line. (The amount of water in storage is increasing.) The amount of water in storage at any time is indicated by the vertical distance between the inflow line and the demand line.

At the end of 1952 the inflow line touches the full reservoir line, indicating that the reservoir is full (point *A*); in 1952 and 1953, however, the demand exceeds the inflow; thus no spilling occurs. Toward the end of 1961 the inflow line crosses the full reservoir line and spilling occurs (point *B*). The spilling continues until 1962 when the demand exceeds the inflow. The water needed to meet the demand is drawn from a full reservoir. To account for the water lost in spilling, the full reservoir line and demand lines are displaced upward as shown. The volume of spill is indicated by the vertical displacement. Spilling occurs again in mid-1965.

There was no time during which the demand was not satisfied. Had there been, the inflow line would have crossed the demand line at the point of failure. The full reservoir and demand lines would have to be displaced downward to indicate the new empty reservoir condition. This graphic technique is called a mass diagram, or Rippl diagram method after an Austrian engineer (1).

The results of the two methods, tabular and graphic, were different because of the respective assumptions concerning the time distribution of inflow and demand. Two extremes were used, neither of which is realistic in most cases. If shorter time periods had been used, much better agreement would be expected.

A disadvantage of the simulation methods described above is that they are based directly on past flows. One may expect future flows to be similar but not identical to those of the past. The sequence is important in this analysis, however, and we can hardly expect the sequence to be the same. Much effort has been made to develop means of synthesizing a statistically similar sequence of flows. Only the simplest method can be described here. More extensive methods are given in Fiering (2), Fiering and Jackson (3), and Hufschmidt and Fiering (4).

The object is to produce a similar sequence of flows. This may be done using the past record and a table of random numbers (see Appendix A). Use two digits of a column of random numbers. Read a random number (say 61). Write down the flow corresponding to that year (279.0 cfs for the Little Blue River). Read the next random number and record the corresponding flow. Continue in this way until the desired number of years of record have been established. Some numbers may be read from the random number table representing years for which there is no recorded flow, the number 31, for example. These are neglected. Some numbers may occur more than once; the corresponding flow is recorded more than once.

The above procedure can be used to synthesize annual flows but should not be used for short time intervals such as months. There is usually a strong correlation between successive flows for short time intervals, and more sophisticated methods are required. Synthesis of flows exhibiting correlation is discussed by Fiering (2), Fiering and Jackson (3), and Hufschmidt and Fiering (4).

NONSEQUENTIAL DROUGHT

The success obtained by using synthetic storms to produce flood estimates of various return periods has led to development of a similar process for developing synthetic drought for a given return period. Synthetic floods were developed starting with maps or graphs depicting rainfall intensity-duration-frequency relationships. This information is seldom available for use in determining low flows; therefore, intensity-duration-frequency relationships of channel flow must be established by the designer. The following steps are used:

1. The minimum duration is selected, say one week.
2. The record of average 7-day flows is searched to determine the smallest 7-day flow. This value is recorded.
3. The record is searched for the next smallest 7-day flow. No day can be used more than once; i.e., the 7-day periods cannot overlap.
4. The process continues until the total record is used.
5. The recorded values are given order numbers M with 1 given to the smallest flow, 2 to the next smallest flow, and so on. A plotting position is assigned by

$$\text{Recurrence interval} = T_e = N/M$$

where N is the number of years of record. The flows are graphed against the recurrence interval. The result is a partial duration series for 7-day low flows.
6. The process is repeated for other durations, say 1 month, 3 months, 6 months, and 1 year. The final result will be a family of discharge,

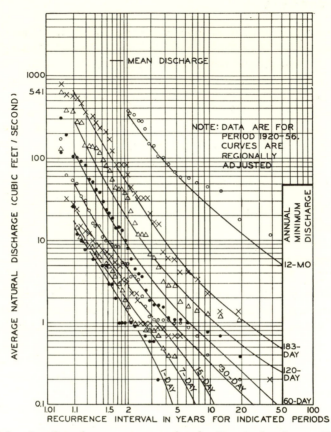

FIG. 8.2. Low-flow frequency curves for the Marais des Cygnes River (5).

duration, and frequency of occurrence of drought. An example of these curves for the Marais des Cygnes River at Ottawa, Kans., is given in Figure 8.2 (5).

A synthetic drought can be constructed for any recurrence interval. For example, a 5-year drought on the Marais des Cygnes River can be expected to indicate the following:

Smallest 7-day flow = 0.26 cfs
Smallest 15-day flow = 0.50 cfs
Smallest 30-day flow = 0.84 cfs
Smallest 60-day flow = 1.40 cfs
Smallest 120-day flow = 2.80 cfs
Smallest 183-day flow = 6.60 cfs
Smallest 1-year flow = 61.0 cfs

This data has some similarity to that read from the maps used in generating a synthetic rainstorm. The synthetic rainstorm was con-

TABLE 8.2. Storage requirements for 5-year drought, Marais des Cygnes River

Duration (days)	Average flow for duration (cfs)	Volume of inflow (cfs-days)	Volume of demand (cfs-days)	Inflow less demand (cfs-days)
7	0.26	1.82	210	−208.18
15	0.50	7.5	450	−442.5
30	0.84	25.2	900	−874.8
60	1.40	84.0	1,800	−2,436.0
120	2.80	336.0	3,600	−3,264.0
183	6.60	1,207.8	5,490	−4,282.2
365	61.00	22,265.0	10,950	11,315.0

structed by rearranging the values into a reasonable sequence. Rearrangement is not usually used with droughts, thus the term *nonsequential drought*.

This drought is used as the flows for a simulation process to determine the required reservoir size. Thus the storage required to maintain 30 cfs flow in the Marais des Cygnes River is established in Table 8.2

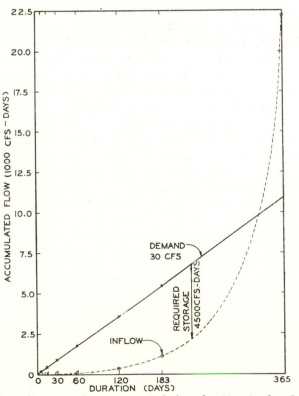

FIG. 8.3. Mass diagram of nonsequential drought, Marais des Cygnes River.

and displayed in Figure 8.3. The maximum difference between the accumulated demand and accumulated inflow is established by constructing
a line parallel to the demand line and locating the point of tangency
with the inflow line. The distance between the inflow and demand lines
at the point of tangency is the required storage, 4500 cfs-days (8920 acre-
feet). This process is attributed to Stall (6).

PROBABILISTIC METHOD (14)

A probabilistic method, credited to Moran (7, 8), can also be used to
analyze reservoir yield. The probability of failing to meet the demand is one of the significant factors resulting from this analysis.

The problem under consideration is the following: A reservoir is
to be constructed on a river and will be used to supply a given quantity
of water. The inflow to the reservoir is a random quantity. What is the
probability that the demand will be met?

To analyze this problem, consider a reservoir with a capacity of K
units (see Fig. 8.4). These units are volume units, and those most convenient should be used. A random inflow enters the reservoir. The inflow has the probability distribution such that the likelihood of i units
entering the reservoir is p_i. If the current content plus the inflow is
greater than the reservoir capacity K, the excess is wasted and is not
considered to satisfy any of the demand. After the inflow period is completed, a withdrawal of M units occurs if possible. If less than M units
are in the reservoir, all that is available is withdrawn. Note that the inflow and demand sequence are identical to those of the tabular method
described previously. The aim of the analysis is to establish the probability of the reservoir being at any particular level and the probability of
not being able to meet the demand.

As an example consider a reservoir with a capacity K of 5 units
subject to a demand M of 2 units. Let P_i be the probability of having i
units in the reservoir initially and P_i' be the probability of having i
units in the reservoir after one inflow withdrawal cycle. The probability
of having 2 units in the reservoir at the end of one cycle would be

$$P_2' = P_3(p_1) + P_2(p_2) + P_1(p_3) + P_0(p_4) \qquad (8.1)$$

That is, P_2' is equal to the sum of the probabilities of having 4 units in
the reservoir before withdrawing 2 units. Similarly,

$$P_1' = P_3(p_0) + P_2(p_1) + P_1(p_2) + P_0(p_3) \qquad (8.2)$$

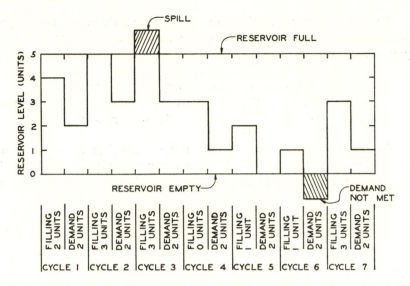

FIG. 8.4.   One possible sequence of inflows into a reservoir of five-unit capacity. The demand is a constant two units. Note that no cycle can end with more than three units in the reservoir.

The probability P_3' is more complex since it occurs due to withdrawal from a full reservoir, and this can occur due to merely filling or filling and wasting. Thus

$$P_3' = P_3(p_2 + p_3 + p_4 + p_5) + P_2(p_3 + p_4 + p_5) + P_1(p_4 + p_5) \\ + P_0(p_5) \tag{8.3}$$

in which p_5 is taken to be the probability of having more than 4 units of inflow.

The probability of ending with an empty reservoir is also more complex, as the reservoir may empty without satisfying the demand. Hence

$$P_0' = P_2(p_0) + P_1(p_1 + p_0) + P_0(p_2 + p_1 + p_0) \tag{8.4}$$

These equations are usually expressed in the form

$$P_3' = P_3(p_2 + p_3 + p_4 + p_5) + P_2(p_3 + p_4 + p_5) + P_1(p_4 + p_5) + P_0(p_5)$$
$$P_2' = P_3(p_1) + P_2(p_2) + P_1(p_3) + P_0(p_4)$$
$$P_1' = P_3(p_0) + P_2(p_1) + P_1(p_2) + P_0(p_3)$$
$$P_0' = P_2(p_0) + P_1(p_1 + p_0) + P_0(p_2 + p_1 + p_0) \tag{8.5}$$

A numerical example is helpful in illustrating the use of these equations. Let the reservoir size be $K = 4$ and the demand be $M = 2$. The equations are

$$P_2' = P_2(p_2 + p_3 + p_4) + P_1(p_3 + p_4) + P_0(p_4)$$
$$P_1' = P_2(p_1) + P_1(p_2) + P_0(p_3) \tag{8.6}$$
$$P_0' = P_2(p_0) + P_1(p_1 + p_0) + P_0(p_2 + p_1 + p_0)$$

For any particular situation the inflow probabilities must be determined from a frequency analysis of the mean annual flow records. However, for this example the inflow probabilities are assumed to be

$$\begin{aligned} p_0 &= 0.1 & p_3 &= 0.3 \\ p_1 &= 0.2 & p_4 &= 0.1 \\ p_2 &= 0.3 & & \end{aligned} \tag{8.7}$$

so that

$$P_2' = 0.7P_2 + 0.4P_1 + 0.1P_0$$
$$P_1' = 0.2P_2 + 0.3P_1 + 0.3P_0 \tag{8.8}$$
$$P_0' = 0.1P_2 + 0.3P_1 + 0.6P_0$$

Consider the reservoir to be empty at $t = 0$; therefore,

$$P_0 = 1, \qquad P_1 = P_2 = 0 \tag{8.9}$$

and, from the equations

$$P_2' = 0.1, \qquad P_1' = 0.3, \qquad P_0' = 0.6 \tag{8.10}$$

That is, there is a 60% chance that at the end of the first time interval the reservoir will remain empty and a 30% chance that it will contain 1 unit.

For the next time interval replace the unprimed P values with the newly calculated primed values,

$$P_2' = 0.7(0.1) + 0.4(0.3) + 0.1(0.6)$$
$$P_1' = 0.2(0.1) + 0.3(0.3) + 0.3(0.6) \tag{8.11}$$
$$P_0' = 0.1(0.1) + 0.3(0.3) + 0.6(0.6)$$

or

$$P_2' = 0.25, \qquad P_1' = 0.29, \qquad P_0' = 0.46 \tag{8.12}$$

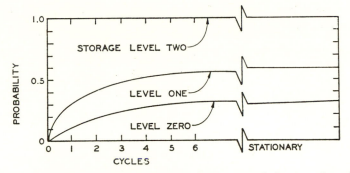

FIG. 8.5. Probability of reservoir storage being at or below a given level at the end of a cycle. Reservoir was initially empty.

At the end of this second time interval the probability that the reservoir will remain empty has improved to 46%. Note that the sum of the probabilities $P_2' + P_1' + P_0'$ is always equal to 1.

This process may be carried on step by step as long as desired. The results of such computations are shown in Figure 8.5. The probability of the reservoir ending a cycle at a given level or less is graphed against time. The lines are drawn only to indicate trends; they do not indicate a continual process during the year. Note that as time progresses, the lines approach the horizontal. This indicates that $P_i' = P_i$, or the distribution is stationary. Figure 8.6 is a graph of a similar set of computations but for a reservoir that is initially full. Note that these curves also approach a stationary distribution, and it is the same distribution approached by the reservoir starting empty. This indicates that later events are not strongly influenced by very early events. Is this reasonable?

The stationary distribution is a significant piece of information and

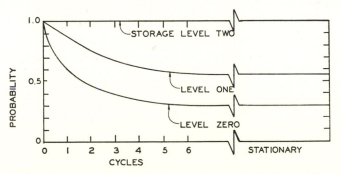

FIG. 8.6. Probability of reservoir storage being at or below a given level at the end of a cycle. Reservoir was initially full.

may be established directly. The condition is that $P_i' = P_i$; therefore, the equations are

$$P_2' = P_2 = 0.7P_2 + 0.4P_1 + 0.1P_0$$
$$P_1' = P_1 = 0.2P_2 + 0.3P_1 + 0.3P_0 \qquad (8.13)$$
$$P_0' = P_0 = 0.1P_2 + 0.3P_1 + 0.6P_0$$

or

$$0 = -0.3P_2 + 0.4P_1 + 0.1P^0$$
$$0 = 0.2P_2 - 0.7P_1 + 0.3P_0 \qquad (8.14)$$
$$0 = 0.1P_2 + 0.3P_1 - 0.4P_0$$

This is a system of three equations with three unknowns, but they are not independent and therefore cannot be solved. A solution can be found if one equation is replaced by the requirement $P_2 + P_1 + P_0 = 1$. That is, the reservoir must contain 0, 1, or 2 units at the end of a cycle. The equations to be solved are now

$$1 = P_2 + P_1 + P_0$$
$$0 = 0.2P_2 - 0.7P_1 + 0.3P_0 \qquad (8.15)$$
$$0 = 0.1P_2 + 0.3P_1 - 0.4P_0$$

The solution to this system of equations is

$$P_2 = 0.442, \qquad P_1 = 0.256, \qquad P_0 = 0.302 \qquad (8.16)$$

This means that after the reservoir has been in operation for some time, the probability that the reservoir will finish the cycle empty is 0.302, that it will contain 1 unit is 0.256, and that it will contain 2 units is 0.442.

The demand will not be met if the reservoir is at level 1 and there is no inflow, or the reservoir is empty and 1 or 0 units enter the reservoir. Thus the likelihood of failure to meet the demand can be evaluated as

$$P_{\text{failure}} = P_0(p_1 + p_0) + P_1(p_0)$$
$$= 0.302(0.3) + 0.256(0.1) \qquad (8.17)$$
$$= 0.116$$

Obviously a large value of P_{failure} would be undesirable. The likelihood of wasting water can be established by applying similar reasoning to the upper portion of the reservoir.

This analytic method is quite easy to use and results in meaningful information not obtained in the simpler mass diagram or simulation approaches. For practical application the reservoir would have to be divided into many more levels, resulting in a larger system of equations to be solved.

The Moran method is based on a model in which the inflow occurs in one season and the demand in the following season. The technique has been extended by many investigators to include more situations. Notable are the efforts of Gould (9), White (10, 11), Dearlove and Harris (12), and Harris (13), some of whose publications are listed in the references.

PROBLEMS

8.1. The following equations were written in order to estimate probabilities of yield for a reservoir:

$$P_5' = P_0(p_8 + p_9 \ldots) + P_1(p_7 + p_8 + p_9 \ldots) + P_2(p_6 + p_7 + p_8 \ldots)$$
$$+ P_3(p_5 + p_6 \ldots) + P_4(p_4 + p_5 \ldots) + P_5(p_3 + p_4 \ldots)$$
$$P_4' = P_0(p_7) + P_1(p_6) + P_2(p_5) + P_3(p_4) + P_4(p_3) + P_5(p_2)$$
$$P_3' = P_0(p_6) + P_1(p_5) + P_2(p_4) + P_3(p_3) + P_4(p_2) + P_5(p_1)$$
$$P_2' = P_0(p_5) + P_1(p_4) + P_2(p_3) + P_3(p_2) + P_4(p_1) + P_5(p_0)$$
$$P_1' = P_0(p_4) + P_1(p_3) + P_2(p_2) + P_3(p_1) + P_4(p_0)$$
$$P_0' = P_0(p_3 + p_2 + p_1 + p_0) + P_1(p_2 + p_1 + p_0) + P_2(p_1 + p_0)$$
$$+ P_3(p_0)$$

What was the capacity of the reservoir being studied? What was the annual demand?

8.2. The Lamine River at Clifton City, Mo., has an average annual flow of 500 cfs. Construction of a reservoir with a capacity of 2500 cfs-years is being planned.

   a. Assuming one storage unit is equal to 500 cfs-years, write out the storage probability equations for this reservoir. Assume a demand of 500 cfs.

   b. Analysis of the record of mean annual flows for the Lamine River yields the following information:

Mean annual flow (cfs)	Probability of exceedance (%)
0	100.0
250	73.0
750	17.0
1250	1.8
1750	0.1
2250	0.0

Determine the probabilities that the annual flow will be within the intervals 0–250, 251–750, 751–1250, 1251–1750, 1751–2250 cfs.

c. Using the probabilities determined in part (b), solve the equations written in part (a) to find the probabilities of storage for the reservoir after one year of operation, assuming the reservoir is empty initially.

d. Determine the stationary values of storage level probabilities, using the probabilities determined in (b).

e. What is the probability that this storage reservoir will not always contain sufficient water to meet the demand?

f. What is the probability that water will be spilled during operation of the proposed reservoir?

8.3. A reservoir has a capacity of 5 units and is operated with an annual demand of 3 units. The inflow probabilities are:

$$p_0 = 0.1, \ p_1 = 0.3, \ p_2 = 0.3, \ p_3 = 0.2, \ p_4 = 0.1, \ p_5 = p_6 = \ldots = 0$$

Find the stationary distribution for the probability of the reservoir being at each of the possible levels.

8.4. Use the average annual flows for the Blue River (Appendix B) to construct a mass diagram. Using the mass diagram, determine the reservoir capacity that would be required to provide a constant annual demand equal to the mean annual flow of the river.

8.5. Construct a mass diagram for the stream having the following monthly discharges during the 2-year period of record. What is the reservoir size required to provide a constant downstream flow of 100 cfs?

Month	Monthly discharge (acre-ft)	
	1961	1962
January	467	83
February	533	250
March	233	350
April	267	432
May	300	316
June	200	233
July	83	133
August	17	100
September	17	100
October	17	33
November	33	67
December	67	17

8.6. Using the annual flows for the Blue River, generate 50 years of similar but different annual flows using random numbers as discussed in the section on simulation methods (first section in this chapter). What is the average annual flow for this new set? How does it compare to the average annual flow for the original set of annual flows?

8.7. Using the set of annual flows generated in Problem 8.6 and assuming a reservoir having a capacity of 408,000 acre-feet, perform a reservoir simulation similar to that shown in Table 8.1. Assume a constant annual demand of 140 cfs.

8.8. The annual flows of a river have a mean of 800 cfs. The standard deviation about the mean is 200 cfs. Assume that the mean flows fit a normal distribution. A reservoir of 3200 cfs-years capacity is being considered for construction on this river.
   a. Write the probability storage equations assuming a constant demand of 600 cfs. Choose your own size of storage unit.
   b. Determine the stationary probabilities of the reservoir being at each of the possible levels.

REFERENCES

1. Rippl, W. The capacity of storage reservoirs for water supply. Proc. Inst. Civil Eng., Vol. 71, 1883.
2. Fiering, M. B. *Streamflow Synthesis.* Harvard Univ. Press, 1967.
3. Fiering, M. B.; and Jackson, B. B. Synthetic streamflows. Am. Geophys. Union, Water Res. Monogr. 1, 1971.
4. Hufschmidt, M. M.; and Fiering, M. B. *Simulation Techniques for Design of Water-Resource Systems.* Harvard Univ. Press, 1966.
5. Furness, L. W. Kansas stream flow characteristics. 2. Low-flow frequency. Kansas Water Resources Board Tech. Rept. No. 2, June 1960.
6. Stall, J. B. Reservoir mass analysis by a low-flow series. *ASCE J. Sanit. Div.*, Vol. 88, No. SA5, Paper 3283, Sept. 1962.
7. Moran, P. A. P. A probability theory of dams and storage systems. *Aust. J. Appl. Sci.*, Vol. 5, 1954.
8. ———. *The Theory of Storage.* Methuen, London, 1959.
9. Gould, B. W. Statistical methods for estimating the design capacity of dams. *J. Inst. Eng.*, Australia, Dec. 1961.
10. White, J. B. A variable season model. Reservoir Yield Symp., Water Res. Assoc., Medmenham, Marlow, Buckinghamshire, England, Sept. 1965.
11. ———. Reservoir yield probability methods. Lecture notes for Hydrology Short Course, Univ. Mo., Columbia, 1969.
12. Dearlove, R. E.; and Harris, R. A. Bivariate matrices for correlated seasons; A pumped storage example. Reservoir Yield Symp., Water Res. Assoc., Medmenham, Marlow, Buckinghamshire, England, Sept. 1965.
13. Harris, R. A. Probability of reservoir yield failure using Moran's steady-state probability method and Gould's probability method. *J. Inst. Water Eng.*, Vol. 19, N4, June 1965.
14. Hjelmfelt, A. T., Jr. Analysis of Reservoir yield. Water and Sewage Works, 1970.

CHAPTER NINE

GROUNDWATER

GROUNDWATER, in general, refers to water as it occurs below the soil surface. Usually, however, groundwater is particularly considered to be the water in the zone of saturation. Seepage into streams, which maintains streamflow in the absence of surface runoff (baseflow), comes from the zone of saturation, as does the water pumped from wells. Water enters the zone of saturation by percolation through the zone of aeration from the ground surface and by seepage from the stream during times of high streamflow.

The intent of this chapter is to provide the fundamentals for analysis of flow in the zone of saturation. Water in the zone of aeration is considered briefly.

WATER IN THE ZONE OF AERATION

Water infiltrating into the ground first enters the root zone of the soil. In this region the pores of the soil contain both air and water. The pressure at the surface is atmospheric, and the water is held in place by capillary or electrochemical forces. This region of the soil system can hold only a limited quantity of water known as the field capacity. When the field capacity is filled (see Fig. 1.3), any additional water added to the system displaces water into the zone of saturation, the groundwater region.

Plants require water in the root zone to maintain life. When the moisture content gets too low the plants wilt; that level is termed the wilting point. The goal of irrigation is to maintain soil moisture at a level above the wilting point but not to unintentionally add water beyond the field capacity. The difference between the field capacity and the wilting point is the maximum quantity of available water in the root

TABLE 9.1. Moisture values for various soils

Soil classification	Field capacity (ft⁻¹)	Permanent wilting point (ft⁻¹)	Available soil moisture at field capacity (ft⁻¹)
Sand	1.2	0.3	0.9
Fine sand	1.4	0.4	1.0
Sandy loam	1.9	0.6	1.3
Fine sandy loam	2.6	0.8	1.8
Loam	3.2	1.2	2.0
Silty loam	3.4	1.4	2.0
Light clay loam	3.6	1.6	2.0
Clay loam	3.8	1.8	2.0
Heavy clay loam	3.9	2.1	1.8
Clay	3.9	2.5	1.4

Source: U.S. Army, Corps of Engineers (1).

zone. Approximate values of field capacity, wilting point, and available water for various soils are given in Table 9.1 (1).

The moisture lost from the root zone may be replaced by precipitation, irrigation, or capillary rise from the upper surface of the groundwater region. The mathematical description of the motion of water in this region is quite complex and is not necessary for most hydrologic work.

WATER IN THE ZONE OF SATURATION

The boundary between the root zone and the zone of saturation is termed the water table or the phreatic line. A well drilled through the soil would fill with water to this line. The pressure in the well increases downward hydrostatically, with atmospheric pressure occurring at the free surface. The above comments apply readily to a continuous soil system with no impermeable layers separating them.

An aquifer is a stratum of the earth that contains and conducts water. Aquifers occur that are unconfined and some exist that are confined, top and bottom, by an impermeable layer. Water moves through the confined aquifers as through a pipe filled with a permeable material. The fluid may be under pressure in these aquifers, and water will rise in a well to a level above the upper impermeable strata. Such an aquifer is termed artesian, and wells that pierce it are called artesian wells.

DARCY'S LAW

In nature a soil or an aquifer is seldom if ever composed of particles of uniform size, shape, and composition. Almost always the

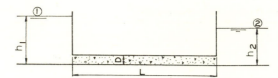

FIG. 9.1. Two reservoirs connected by a pipe filled with a permeable material.

presence of lenses of relatively impermeable material, rocks, or abrupt discontinuities complicate the situation.

Water flowing as groundwater moves through tortuous paths formed by interconnected voids between soil particles. Because the resistance to flow is high, the velocity with which the water moves is slow and as a result the flow is laminar in nature (2). As an example, consider flow occurring between two reservoirs connected by a pipe filled with a uniformly graded porous material as shown in Figure 9.1. Writing the energy equation from the surface of the first reservoir to the surface of the second yields

$$\frac{V_1^2}{2g} + \frac{P_1}{\gamma} + Z_1 = \frac{V_2^2}{2g} + \frac{P_2}{\gamma} + Z_2 + f\,\frac{L}{D}\,\frac{V^2}{2g} \qquad (9.1)$$

in which

$$V = \text{velocity}$$
$$P = \text{pressure}$$
$$Z = \text{elevation}$$
$$g = \text{gravitational acceleration}$$
$$f = \text{Weisbach resistance coefficient}$$
$$L = \text{length of pipe}$$
$$D = \text{pipe diameter}$$
$$\gamma = \text{specific weight of the fluid}$$

Since V_1 and V_2 are normally quite small, we neglect smaller terms containing velocity squared. Because $P_1 = P_2 = $ atmospheric pressure, Eq. (9.1) becomes

$$Z_1 - Z_2 = h_1 - h_2 = f(L/D)\,(V^2/2g) \qquad (9.2)$$

or

$$V^2 = (2gD/f)\,[h_1 - h_2)/L] = -(2gD/f)\,(\Delta h/L) \qquad (9.3)$$

where $\Delta h = h_1 - h_2$. Studies have shown that for the very small velocities normally encountered, the resistance coefficient f is inversely proportional to V. The quantity $2gD/f$ is usually defined as

$$2gD/f = KV \qquad (9.4)$$

where K is called the permeability coefficient and is known to be a function of relative grain size and shape (3). Permeability is thus the opposite of resistance, and low resistance to flow implies relatively high permeability. Substituting Eq. (9.4) into Eq. (9.2) and reducing to a differential length dL yields

$$V = -K \frac{dh}{dL} \qquad (9.5)$$

which is one form of Darcy's law (4). The minus sign in Eq. (9.5) simply indicates that the flow moves in the direction of decreasing h. Note that K has the dimensions of a velocity in Eq. (9.5).

Certain limitations must be recognized when using Eq. (9.5). The velocity is not the true velocity of the fluid as it winds its way in, out, and around the porous material but is a fictitious velocity given by dividing the flow rate Q by the cross-sectional area of the pipe, the total area of both pores, and solids projected on a plane normal to the pipe center line. In the preceding example V is the flow rate divided by the cross-sectional area of the pipe. Equation (9.5) is valid only for small velocities and becomes less and less correct as the velocity increases.

Extensive developments of Darcy's law and discussions of its limitations and the concept of permeability are given by Muskat (4) and Scheidegger (5).

Some representative values of the permeability K are given in Table 9.2 (6). Several different forms of Darcy's equation and units are in

TABLE 9.2. Representative permeabilities for various materials

Material	Permeability (K)	
	gal/day/ft²	(ft/sec) $\times 10^6$
Clay, silt	0.001–2	0.0015–3
Sand	100–3,000	150–4,700
Gravel	1,000–15,000	1,500–23,000
Gravel and sand	200–5,000	300–8,000
Sandstone	0.1–50	0.15–80
Shale	0.00001–0.1	0.000015–0.15

Source: Walton (2).

common use, and care is necessary when comparing values from other sources. One of the standard devices for measuring permeability involves a device that could be closely represented by Figure 9.1. A sample of the material of interest is placed in a particular fashion in the conduit. Rate of flow between the reservoirs is measured as well as Δh, and K is then computed from Eq. (9.5).

In Table 9.2 the permeability is seen to be a function of material type. Gravel, with its large particle size and thus correspondingly large interstitial openings, provides relatively little resistance to flow, while the densely packed particles in shale provide relatively great resistance.

TWO-DIMENSIONAL FLOWS

Some examples of two-dimensional flows will illustrate the application of Darcy's law.

FLOW TO A TRENCH THROUGH A CONFINED AQUIFER

A porous material is confined between two impermeable layers as shown in Figure 9.2. Flow from the river maintains the porous material in saturated condition. A long trench is dug through the porous material, and water is pumped from the trench at such a rate that the depth of the water in the trench remains steadily at h_2. How much water per foot of trench must be pumped out? Note that the trench fully penetrates the aquifer. This avoids the problem of flow through the bottom of the trench. Fully penetrating trenches and wells will be used throughout this chapter. Note also that the aquifer is assumed to be fully saturated throughout.

This problem is similar to the two reservoirs with the connecting pipe discussed earlier except that the pipe is very wide and rectangular. The cross-sectional area of the aquifer is D for a unit width, and the flow rate through this unit width of aquifer is given by

$$Q = VA = VD \tag{9.6}$$

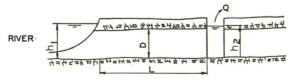

FIG. 9.2.   Flow to a trench through a confined aquifer.

The velocity is given by Eq. (9.5), and thus the expression for Q becomes

$$Q = -KD \frac{dh}{dL}$$

We assume that the piezometric head varies linearly from h_1 to h_2 so that

$$\frac{dh}{dL} = \frac{h_2 - h_1}{L}$$

Thus

$$Q = [(h_1 - h_2)/L]KD \tag{9.7}$$

FLOW TO A TRENCH THROUGH A PARTIALLY CONFINED AQUIFER

Quite often wells are used to dewater a construction area. Thus Figure 9.2 might correspond to a problem condition, and it is desirable to pump enough water from the trench to partially dry out material near it so that the water level h_2 is less than the aquifer thickness D. Such a practice is common in construction sites where high water tables are encountered. Pumping down to h_2 in this case leaves the groundwater near the trench unconfined by the upper impermeable layer, although that near the river remains confined. However, we do not know at the outset what part of length L is saturated. Figure 9.3 illustrates this example. The same situation arises if flow into a stream is considered and the level of the stream falls below the top of the aquifer.

To establish the rate per foot of trench length at which water must be pumped, it is helpful to separate the problem into two separate parts. Consider first the confined flow occurring between the river and the point where the flow becomes unconfined, at the unknown distance B. This flow is given by

$$Q = [(h_1 - D)/B]KD \tag{9.8}$$

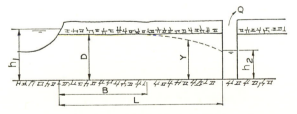

FIG. 9.3.   Flow to a trench through a partially confined aquifer.

as determined in our previous example. The second problem considers the unconfined flow. Again the flow is given by

$$Q = -K \frac{dh}{dL} A \qquad (9.9)$$

The area in this case is not a constant but varies along the aquifer with distance x as measured in Figure 9.3. Note also that y, the elevation of the water surface, is the same as the vertical dimension of the area so that the cross-sectional area is $A = y$. The flow rate per unit width is given by

$$Q = -K \frac{dy}{dL} y \qquad (9.10)$$

This is a differential equation whose solution can be expressed as the integral

$$Q \int_B^L dL = -K \int_D^{h_2} y \, dy \qquad (9.11)$$

which results in

$$Q = K \frac{D^2 - h_2^2}{2(L - B)} \qquad (9.12)$$

Because the flow that passes through the confined portion must also pass through the unconfined portion, the two Q values are the same. The two equations, Eq. (9.8) and Eq. (9.12), for Q may be solved for the two unknowns Q and B:

$$Q = (K/2L) [2D(h_1 - D) + (D^2 - h_2^2)]$$
$$B = \frac{2LD(h_1 - D)}{2D(h_1 - D) + (D^2 - h_2^2)} \qquad (9.13)$$

In the solution to this example we have neglected any variation in K with velocity, and we have tacitly assumed that Eq. (9.2), written for a uniform flow, is applicable to a converging flow. These assumptions are warranted as long as velocities remain quite small.

Groundwater occurs in both confined and unconfined conditions, as illustrated in the two hypothetical examples. Accordingly, the aquifer or saturated region is defined as a confined or unconfined aquifer. In a confined aquifer the water is under pressure, as water in a pipe. If a well is drilled into a confined aquifer, water will rise in the well hole above the level of confinement. In fact, it will rise to the level

$$y = P/\gamma \tag{9.14}$$

where

$\gamma =$ specific weight of the fluid (62.4 lb/cu ft for water)
$P =$ pressure in pounds/square foot
$y =$ feet of rise above point where P is measured

When Darcy's law is used for confined aquifers, the term h is

$$h = Z + (P/\gamma) \tag{9.15}$$

where

$Z =$ the elevation above some arbitrary datum
$P =$ the pressure at the elevation Z

In arriving at Eq. (9.2), P_1 and P_2 were at the corresponding free surfaces and both were equal to atmospheric pressure. Pressure at the surface of any unconfined flow is likewise assumed to be atmospheric.

STEADY-STATE WELL FLOWS

An example of analysis of flow into a well will be given as an additional illustration of the application of Darcy's law. In this particular example the flow to the well is through an unconfined aquifer that is homogeneous in all directions and is limited on the bottom by an impermeable layer. The previous flows were two-dimensional in a vertical plane. The flow is steady; i.e., no conditions vary with time. This flow is likewise two-dimensional but in a radial plane. Obviously, such steady flow cannot occur unless the water being drawn from the well is continuously being replaced uniformly around the well at a great distance. Figure 9.4 illustrates the flow.

Water is pumped from the well of radius r_w at a rate Q. The flow is unconfined at the top. What will be the elevation h_w of the water surface in the well?

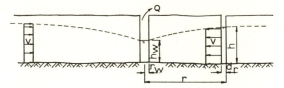

FIG. 9.4. Flow to a well in an unconfined aquifer.

The basic equation is again Darcy's law:

$$Q = -K \frac{dh}{dL} A$$

The flow to the well must pass radially with velocity V through the sides of a cylinder of height h, radius r, and thickness dr (Fig. 9.4).

The flow is moving radially inward toward the well from all sides. Thus, Eq. (9.9) becomes

$$Q = \left(K \frac{dh}{dr} \right) [(2\pi r)h] \qquad (9.16)$$

Note that $dL = -dr$ because the flow is in the direction of decreasing r. The solution to this differential equation can be expressed in the integral form

$$Q \int_{r_w}^{R} \frac{dr}{r} = 2\pi K \int_{h_w}^{h_1} h \, dh \qquad (9.17)$$

which integrates to give

$$h_w = [h_1^2 - (Q/\pi K) \ln (R/r_w)]^{1/2} \qquad (9.18)$$

The quantity $h_1 - h_w$ is the amount that the pumping has lowered the water table and is termed the *drawdown*. In this solution no energy loss has been considered where the water enters the well from the aquifer. In actual cases a well screen is used there, and flow through it must be considered (7).

UNSTEADY-FLOW WELL PROBLEMS

Although flow into a well is frequently treated as a steady-flow problem (conditions do not change with time), most cases are actually unsteady (conditions vary with time). Groundwater in most cases is recharged through infiltration during the time precipitation occurs and for a limited time thereafter. When water is pumped from the groundwater table at a steady rate, the total amount of water stored in the groundwater is reduced but is replenished periodically by storms. Thus the free surface in an unconfined aquifer draws down further as pumping continues. Exceptions occur when a river or lake recharges the aquifer directly. The unsteady flow solution for a confined well was derived by Theis (8). The application of this solution will be demonstrated. For the derivation the reader is referred to Theis (8) or advanced texts devoted to groundwater, e.g., Davis and DeWiest (2) or Walton (6).

NONEQUILIBRIUM FLOW TO A WELL IN A CONFINED AQUIFER

Water is pumped from the well at a constant rate Q. The flow out of the well exceeds the flow into the system; as a result, water is taken out of storage, and the drawdown increases with time. As the drawdown increases, the rate of flow into the system increases until equilibrium is achieved. This of course assumes that flow enters the aquifer at a rate equal to Q at a great distance from the well. That assumption may not always be true. In fact, many groundwater basins in the United States, particularly in the Southwest, have been seriously depleted by continuous pumping.

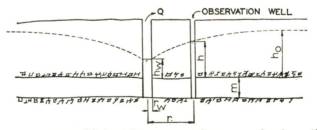

FIG. 9.5.   Nonequilibrium flow to a well in a confined aquifer.

Figure 9.5 shows a confined aquifer that will be analyzed for unsteady flow. Before pumping starts, the piezometric head forms a level plane at elevation h_0. When pumping starts, the surface of revolution formed by the piezometric head drops rapidly near the well and less rapidly at a distance from the well. We would like to establish the position of this surface as time passes.

The solution of the partial differential equation describing the unsteady system is in the form of an exponential integral and may be expressed as (8)

$$h_0 - h = [114.6Q/T]W(u) \tag{9.19}$$

in which

$h_0 - h$ = drawdown (feet)
Q = discharge (gallons/minute)
T = coefficient of transmissibility (gallons/day/foot)
$W(u)$ = a well function

The coefficient of transmissibility T is the product of the permeability K and the thickness of the aquifer m, and has the dimensions of gallons/day/foot. The well function $W(u)$ is the exponential integral, which cannot be evaluated in terms of simple functions. Values for various arguments u are given in Table 9.3 (9). The argument is given by

$$u = [1.87S/T](r^2/t) \qquad (9.20)$$

in which

r = distance from the pumped well to the observation point (feet)
t = time after pumping started (days)
S = a coefficient of storage

The coefficient of storage is related to the compressibility of the system and may be considered to be the volume of water removed from a 1-foot square vertical column in the aquifer when the head p/γ is lowered 1 foot.

Equations (9.19) and (9.20) can be used in two ways. The drawdown, and hence the pumping lift, can be determined as a function of time for a well where both transmissibility and storage coefficient are known. This is found by direct substitution, using the well radius as the distance to the observation point. Inversely, the aquifer properties T and S can also be determined by pumping a well at a constant rate Q and measuring the drawdown at various times.

The aquifer properties cannot be established directly, but a graphic technique is available. To approach this procedure, note that in Eq. (9.19) and Eq. (9.20) the terms in the brackets are constant for a particular well and can be treated as constants of proportionality. As a result, a graph of $W(u)$ vs. u should have the same shape as a graph of $h_0 - h$ vs. r^2/t. Only the scales will differ between the two graphs, the difference resulting from the constants of proportionality. The graphic procedure is to graph $W(u)$ vs. u, using Table 9.3 as shown in Figure 9.6.

Using the measured values, another graph is constructed of $h_0 - h$ vs. r^2/t as shown in Figure 9.7. One of the graphs is drawn on transparent paper, and both are drawn on log-log graph paper. The two graphs are superimposed (as shown in Fig. 9.8) and translated, keeping the coordinates parallel, until the curves coincide as well as possible.

TABLE 9.3. Values of W(u) for values of u

u	1.0	2.0	3.0	4.0	5.0	6.0	7.0	8.0	9.0
$\times 1$	0.219	0.049	0.013	0.0038	0.0011	0.00036	0.00012	0.000038	0.00
$\times 10^{-1}$	1.82	1.22	0.91	0.70	0.56	0.45	0.37	0.31	0.26
$\times 10^{-2}$	4.04	3.35	2.96	2.68	2.47	2.30	2.15	2.03	1.92
$\times 10^{-3}$	6.33	5.64	5.23	4.95	4.73	4.54	4.39	4.26	4.14
$\times 10^{-4}$	8.63	7.94	7.53	7.25	7.02	6.84	6.69	6.55	6.44
$\times 10^{-5}$	10.94	10.24	9.84	9.55	9.33	9.14	8.99	8.86	8.74
$\times 10^{-6}$	13.24	12.55	12.14	11.85	11.63	11.45	11.29	11.16	11.04
$\times 10^{-7}$	15.54	14.85	14.44	14.15	13.93	13.75	13.60	13.46	13.34
$\times 10^{-8}$	17.84	17.15	16.74	16.46	16.23	16.05	15.90	15.76	15.65
$\times 10^{-9}$	20.15	19.45	19.05	18.76	18.54	18.35	18.20	18.07	17.95
$\times 10^{-10}$	22.45	21.76	21.35	21.06	20.84	20.66	20.50	20.37	20.25
$\times 10^{-11}$	24.75	24.06	23.65	23.36	23.14	22.96	22.81	22.67	22.55
$\times 10^{-12}$	27.05	26.36	25.96	25.67	25.44	25.26	25.11	24.97	24.86
$\times 10^{-13}$	29.36	28.66	28.26	27.97	27.75	27.56	27.41	27.28	27.16
$\times 10^{-14}$	31.66	30.97	30.56	30.27	30.05	29.87	29.71	29.58	29.46
$\times 10^{-15}$	33.96	33.27	32.86	32.58	32.35	32.17	32.02	31.88	31.76

Source: Wenzel (9).

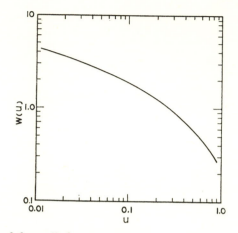

FIG. 9.6.   Graph of the well function, W(u) vs. u, frequently called the type curve.

A point on the coincident curves is selected and $W(u)$, $h_0 - h$, u, and r^2/t are recorded for that point. The constants of proportionality are determined as

$$(h_0 - h)/W(u) = [114.6Q/T] \qquad\qquad (9.21)$$

or

$$T = 114.6Q[W(u)/(h_0 - h)] \qquad\qquad (9.22)$$

and

$$u/(r^2/t) = 1.87S/T \qquad\qquad (9.23)$$

or

$$S = [T/1.87][u/(r^2/t)] \qquad\qquad (9.24)$$

The preceding development is limited to fully penetrating wells in confined aquifers. Similar developments for partially penetrating wells, wells in leaky confined aquifers, and wells in unconfined aquifers can be found in books devoted to groundwater hydrology such as Muskat (4), Davis and DeWiest (2), Walton (6), Harr (10), or Polubarinova-Kochina (11). Details of well tests for more complex situations are given in the U.S. Geological Survey water supply papers (7, 12).

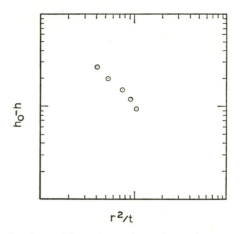

FIG. 9.7. Graph of values of $h_o - h$ vs. r^2/t, using values measured in the field.

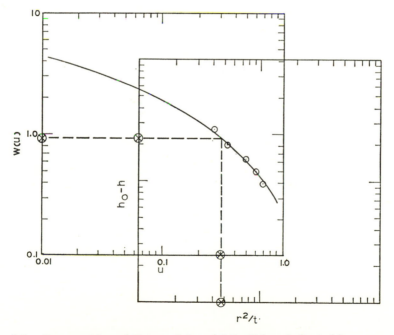

FIG. 9.8. Superposition of Figures 9.6 and 9.7. Values indicated by x are used in Eq. (9.30) and Eq. (9.41) to determine T and S.

PROBLEMS

9.1. Water travels from the river source to a trench where the water level is maintained at h_2 by pumping (Fig. P.9.1). Derive an equation for the rate of water per foot of trench that must be removed to maintain the level in the trench. The permeability of the strata is K.

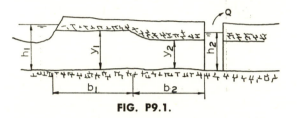

FIG. P9.1.

9.2. Solve Problem 9.1 for the situation shown in Figure P9.2.

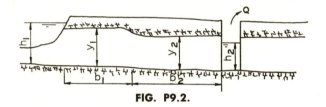

FIG. P9.2.

9.3. Water flows to a well through a confined aquifer. A test well at radius r_2 indicates that the hydrostatic pressure at that location is h (Fig. P9.3). Water is withdrawn from the well at the rate Q. Derive an equation for the equilibrium water elevation h_w in the well if the permeability is K.

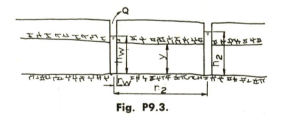

Fig. P9.3.

9.4. Solve Problem 9.3 for the flow situation depicted in Figure P9.4.

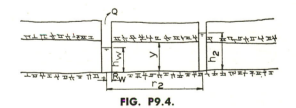

FIG. P9.4.

9.5. A well was pumped continuously for 48 hours at a rate of 540 gallons/minute. A series of observation wells, located at various distances from the pumped well were checked; the results are listed below. Determine the transmissibility and coefficient of storage for this well. (Adapted from [7].)

Distance from pumped well (ft)	Drawdown (ft)
49.7	2.94
170.0	1.43
270	0.84
430	0.45
625	0.21
805	0.11
940	0.09

9.6. A new well was pumped continuously at a rate of 400 gallons/minute. The drawdown at a single observation well located 200 feet from the pumped well was measured; the results are tabulated below. Determine the transmissibility and coefficient of storage for the aquifer.

Time (days)	Drawdown (ft)
0.075	1.00
0.15	2.57
0.25	4.20
0.375	5.60
0.75	8.35

9.7. A 12-inch diameter well is to be located in an aquifer with a transmissibility of 8000 gallons/day/foot, and a coefficient of storage of 0.07. The initial water level in the well was 60 feet below ground surface. If 400 gallons/minute are pumped from the well continuously, what will be the pumping lift after 1 year and after 5 years?

REFERENCES

1. U.S. Army, Corps of Engineers. Snow hydrology. North Pacific Div., Portland, Oreg., June 1956.
2. Davis, S. N.; and DeWiest, R. J. M. *Hydrogeology*. Wiley, 1966.
3. Rouse, H., ed. *Engineering Hydraulics*. Wiley, 1950.
4. Muskat, M. *The Flow of Homogeneous Fluids through Porous Media*. J. W. Edwards, 1946.
5. Scheidegger, A. E. *The Physics of Flow through Porous Media*. Macmillan, 1960.
6. Walton, W. C. *Groundwater Resource Evaluation*. McGraw-Hill, 1970.

7. U.S. Geological Survey. Methods of determinng permeability, transmissibility and drawdown. Water Supply Paper 1536-I, Washington, D.C., 1963.

8. Theis, C. V. Relation between the lowering of the piezometric surface and the rate and duration of discharge of a well using groundwater storage. Trans. Am. Geophys. Union, Part 2, pp. 519–24, 1935.

9. Wenzel, L. K. Method for determining permeability of water bearing materials with special reference to discharging-well methods. U.S. Geological Survey Water Supply Paper 887, Washington, D.C., 1943.

10. Harr, M. E. *Groundwater and Seepage.* McGraw-Hill, 1962.

11. Polubarinova-Kochina, P. Ya. (trans. R. J. M. DeWiest). *Theory of Groundwater Movement.* Princeton Univ. Press, 1962.

12. U.S. Geological Survey. Theory of aquifer tests. U.S. Geological Survey Water Supply Paper 1536-E, Washington, D.C., 1962.

CHAPTER TEN

EVAPORATION

RAINFALL that does not run off either as surface or subsurface flow is referred to as water loss. The hydrologic cycle in Chapter 1 indicated this loss as evapotranspiration. In actuality, evapotranspiration includes evaporation from water surfaces or moist solid surfaces (such as earth or pavement) and transpiration from plants. Water is also incorporated into the physical structure of plants in the growing process and in the case of ice or snow may sublimate, passing directly from a solid to a vapor state. The importance of these losses depends upon the relative value of potential evapotranspiration and annual precipitation. Thus in central Missouri where average annual precipitation (35 inches) approximately equals average annual evaporation from a reservoir surface, there is little or no need to consider evaporation in forming a water budget. In central Arizona, however, average annual evaporation and precipitation are 70 and 10 inches respectively, producing a net loss of approximately 60 inches and making it necessary to give close attention to potential evaporation. Figure 10.1 shows average annual lake evaporation for the United States in inches/year (1).

THE PROCESS OF EVAPORATION

If thermal energy is added to a body of water with a free surface, the kinetic energy of the molecules is increased to the extent that some water molecules at the surface can overcome their surrounding cohesive bonds and are able to escape across the water-air interface. As the molecule of water passes from the liquid to vapor state, it absorbs heat energy, thus cooling the water left behind. The amount of heat absorbed by a unit mass of water transformed from liquid to vapor at constant temperature is known as the latent heat of evaporation L_e and is

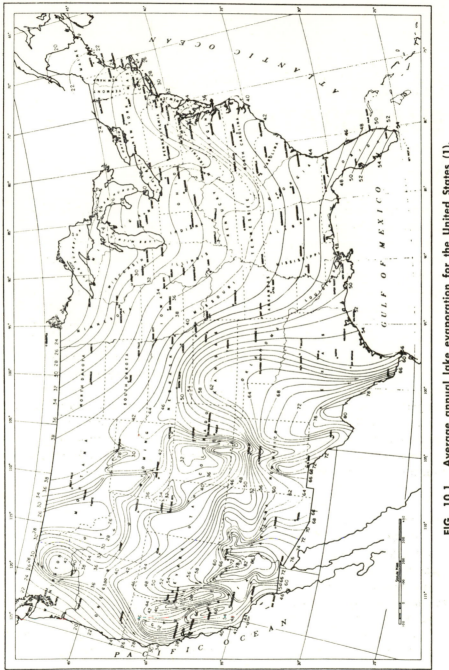

FIG. 10.1. Average annual lake evaporation for the United States (1).

equal to 597.3 calories/gm at 0° C (2). In a closed system, evaporation can continue until the air just above the water surface becomes saturated with water vapor. Condensation must then occur before further evaporation can take place. Condensation and evaporation then occur simultaneously at equal rates, resulting in an equilibrium condition.

In natural situations such as a stream, lake, or reservoir molecules of water vapor are convected away from the water surface by molecular diffusion and wind in the essentially infinite atmosphere. Thus a gradient is formed, with the water vapor content decreasing away from the water surface.

Because of the processes involved, two basic approaches to analyzing evaporation have developed, the diffusion method and the energy balance method. In the diffusion method a formulation is made of the mass transfer process by which vapor is removed from the water surface. In the energy balance method an accounting of energy is made over a period of time, resulting in an average evaporation rate.

Consideration of the process indicates that the rate of evaporation from a water surface depends upon the air temperature, pressure, and relative humidity; water temperature; and the rate at which vapor molecules are carried away from the water surface by molecular diffusion or convection or both. Assuming that convection is a function only of wind speed and boundary layer shape and that molecular diffusion is a function of the previously named physical properties of air, the rate of evaporation is thus a function of at least six variables. Geometry of the reservoir is a gross effect that must also play an important local role (3).

ESTIMATION OF EVAPORATION

Water resource planning will frequently require estimation of at least annual expected evaporation. In the case of a reservoir or long canal in the desert Southwest the designed capacity could be seriously in error if allowance were not made for evaporation. Several methods for use in estimating evaporation have evolved. Because each method represents some degree of simplification of the actual process, the estimations resulting from the various methods will differ, sometimes considerably.

EVAPORATION PANS

Evaporation measurements are made at all U.S. Weather Bureau stations with an evaporation pan known as the *U.S. Weather Bureau class A land pan*. This pan is 4 feet in diameter and 10 inches deep and is supported 6 inches above the ground on a wooden frame. For a variety of reasons evaporation occurs more rapidly from a pan than from a larger lake or reservoir. Therefore, a pan coefficient has been recommended for converting pan evaporation to lake or reservoir equivalents (4):

$$E_R = C_p E_p \tag{10.1}$$

where

E_R = depth of evaporation from the reservoir
E_p = depth of evaporation from the pan
C_p = the pan coefficient

Observed pan coefficients apparently range from 0.6 to 0.8 with an average value of 0.7 being recommended. For a large project, evaporation measurement stations would be established in the local area along with stream gages and weather stations for collection of data to be used in planning and design. However, regular observations of pan evaporation are summarized monthly by the U.S. Weather Bureau (5).

DALTON'S LAW DIFFUSION METHOD

John Dalton first pointed out in 1802 that the rate of evaporation from a water surface will be proportional to the difference between saturated vapor pressure at water temperature and the aqueous vapor pressure of the air (6). In equation form Dalton's law is

$$E = C(e_s - e_a) \tag{10.2}$$

where

E = the evaporation in inches/day
e_s = saturation vapor pressure at the water surface temperature T_s
e_a = aqueous vapor pressure of the air (inches of mercury)

C is a coefficient whose value depends on factors such as wind speed that are unaccounted for. Many equations having the general form of Dalton's law have evolved over the years. Those more commonly used are as follows:

Fitzgerald (1886) (7):

$$E_{\text{day}} = (0.40 + 2V)(e_s - e_a) \tag{10.3}$$

Russell (1888) (8):

$$E_{\text{day}} = [(1.96P_W + 43.88)/P_a](e_s - e_a) \tag{10.4}$$

Horton (1917) (9):

$$E_{\text{day}} = 0.4[(2 - e^{-0.2V})(e_s - e_a)] \tag{10.5}$$

Rohwer (1931) (10):

$$E_{\text{day}} = (1.13 - 0.0143P_a)(0.44 + 0.118V)(e_s - e_a) \tag{10.6}$$

Meyer (1915) (11):

$$E_{\text{day}} = 0.5(e_s - e_a)(1 + 0.1V_{30}) \tag{10.7}$$

Lake Hefner (1954) (12):

$$E_{\text{day}} = 0.06V_{30}(e_s - e_a) \tag{10.8}$$

Lake Hefner (1954) (12):

$$E_{\text{day}} = (0.068 + 0.059V_{13})(e_s - e_a) \tag{10.9}$$

Lake Mead (1958) (13):

$$E_{\text{day}} = .072V_{30}(e_s - e_a)[1 - 0.03\ (T_a - T_w)] \tag{10.10}$$

In the preceding equations the following definitions apply:

E_{day} = depth of evaporation (inches/day)
h = relative humidity (decimal)
V = wind velocity (location not defined)
P_a = atmospheric pressure (inches of mercury)
e_a = actual vapor pressure of air (inches of mercury)
e_s = saturation vapor pressure of air near water surface (inches of mercury)
P_w = vapor pressure near the water
V_{30} = wind velocity at 30 feet above ground (miles/hour)
V_{13} = wind velocity at 13 feet above ground (miles/hour)
T_a = average air temperature (°C)
T_w = average surface water temperature (°C)

In the Meyer equation, Eq. (10.7), the coefficient 0.5 is used for small shallow lakes, while 0.37 should be used for large deep lakes (11). All the equations are of the Dalton or diffusion method form. Comprehensive measurements made cooperatively by several government agencies on Lake Hefner in Oklahoma and Lake Mead in Arizona gave rise to Eqs. (10.8), (10.9), (10.10), and (10.14). Subsequent analysis of the Lake Hefner data gave rise to still another equation (13).

$$E_{day} = C_p(0.37 + 0.0041L_6)(e_{s5} - e_{a5})^{0.88} \qquad (10.11)$$

where e_{s5} and e_{a5} are as in Eq. (10.3) but are measured 5 feet above the ground surface and

L_6 = the total wind movement (miles/day) measured 6 inches above the evaporation pan rim

C_p = the pan coefficient of Eq. (10.1)

In all the empirical equations presented (Eqs. 10.3–10.11), average values of all quantities should be used. In practice, however, average values are seldom known. Considerable variation should therefore be expected of the evaporation depth predicted by use of the empirical relationships. When attempting to predict evaporation, it is prudent to make use of more than one equation and to consider the range of resulting predictions.

As might be gathered from considering the empirical equations (Eqs. 10.3–10.11), wind velocity and gradient in vapor pressure appear to be the important variables. Although the other variables listed in the section on the process of evaporation at the beginning of this chapter can also be expected to be of importance, for particular equations their individual effect is difficult if not impossible to quantify. Thus Russell's equation, Eq. (10.4), although it includes the barometric pressure, could hardly be expected to give reasonable predictions for a reservoir well-exposed to wind.

ENERGY BALANCE METHOD

The concept of the energy balance method is simply stated as: If one can determine the change in energy level of the reservoir due to evaporation over a period of time t, the total volume of evaporation that took place can be computed by dividing that energy by the latent heat of vaporization. In equation form the energy balance is

$$Q_E = Q_S + Q_A - Q_B - Q_N - Q_R \qquad (10.12)$$

where

Q_E = rate of utilization of energy through evaporation
Q_S = net rate of change of energy due to short-wave solar radiation
Q_A = net rate at which energy is transported into and out of the reservoir by inflow, outflow, precipitation, surface runoff, seepage, bank flow, and any other flow of water
Q_B = net rate of outgoing long-wave radiation
Q_N = rate of conduction and convection of energy from the water surface to the atmosphere
Q_R = net rate of change of energy storage within the reservoir

As implied in the concept of latent heat,

$$Q_E = \rho_W L_E E A_R / \Delta t \qquad (10.13)$$

where

$$\rho_W = \text{density of water (lb-sec}^2/\text{ft}^4)$$
$$E = \text{depth of evaporation (ft)}$$
$$A_R = \text{reservoir surface area (ft}^2)$$
$$\Delta t = \text{elapsed time}$$

Q_N and Q_E are closely related because the diffusion of both heat energy and water vapor depend upon convection and conduction. Thus warm wind blowing over a reservoir warms the water and in turn is cooled. For average wind conditions (15, 16)

$$Q_N = 6.1 \times 10^{-4} Q_E P_a [(T_w - T)/(e_s - e_a)] \qquad (10.14)$$

Equation (10.14) and Eq. (10.13) combine to yield

$$E = \frac{Q_S - Q_B + Q_A - Q_R)\Delta t}{\rho_W L_E A_R (1 + 6.1 \times 10^{-4} P_a)[(T_w - T)/(e_s - e_a)]} \qquad (10.15)$$

To use Eq. (10.15) for the prediction of evaporation requires having information available on the energy terms in the numerator. All terms in the denominator of Eq. (10.15) represent average physical properties of the water or air during the time period Δt, and as such may be available for the particular location in U.S. Weather Bureau records. As in any other numerical calculation the maximum size of Δt should be limited so that E will most closely represent the true value. A Δt of 1 month is convenient to use with monthly weather summaries. Lake Hefner studies showed that a good approximation to actual evaporation is obtained if a Δt of at least 7 days is used because changes in energy storage cannot be measured with sufficient accuracy for smaller periods of time (17).

The Lake Hefner studies indicated that the energy balance method can be expected to yield accurate evaporation estimates only if the energy quantities Q_S and Q_B are measured at the site of interest. Thus the energy balance method is inconvenient for use in planning. In addition, it is difficult at best to estimate the Q_A term since this involves seepage flow into and out of the reservoir, both of which are quite difficult to measure or estimate with confidence.

For a lake with an appreciable temperature stratification determination of the total energy contained in the reservoir, $\rho_W V T_w$ (V is the volume, ρ_W and T_w are averages), will require temperature measurement as a function of depth and subsequent integration over the water volume.

EVAPORATION STUDY

Lake MacBride lies slightly north of Iowa City, Iowa, and is used primarily for recreation. The owners wish to know if it is feasible to increase the height of the dam to form a larger lake. As a first step in answering that question a water budget must be constructed. The following information is pertinent:

Surface area of the proposed reservoir = 960 acres (with water surface = elevation 712 feet)
Change in storage capacity per foot of depth = 920 acre-feet/foot
Consider the lake to be full (water surface elevation = 712 feet) on April 1

The climatological data shown in Table 10.1 have been obtained.

In order to compare results of various equations, evaporation will be computed using the Rohwer, Eq. (10.6); the Meyer, Eq. (10.7); the Lake Hefner, Eq. (10.8); and the pan, Eq. (10.1), equations. A pan coefficient of 0.7 is assumed.

The water surface will be assumed to be at the same temperature as the air. In the spring months this will produce a larger than actual evaporation. Wind velocity can be calculated from the power law equation for velocity distribution in a turbulent boundary layer.

$$u/u_0 = (y/y_0)^{1/7} \tag{10.16}$$

where

$$u = \text{velocity}$$
$$y = \text{distance from the ground}$$

TABLE 10.1. Climatological data for Lake MacBride for the open season of 1940

Month	Total inflow (cfs-days)	Average air temperature (°F)	Precipitation (in.)	Pan evaporation (in.)	Mean wind speed at 9 in. above ground (mph)	Relative humidity (%)	Barometric pressure (in. Hg)
Apr.	296	48.0	3.03	4.60	4.0	64	29.96
May	110	58.5	2.11	5.76	3.1	62	29.91
June	107	71.9	2.76	7.76	2.0	61	29.89
July	3	76.3	2.78	9.36	1.7	58	30.00
Aug.	16	71.3	4.11	5.58	1.3	75	29.99
Sept.	26	64.9	2.28	4.98	1.4	64	30.08
Oct.	3	57.2	2.19	3.80	1.9	59	30.01

TABLE 10.2. Evaporation computations

Month	Air temperature (°F)	P_W (in. Hg)	P_A (in. Hg)	PA_{sat} (in. Hg)	Humidity h (%)	Barometric pressure, P_B (in. Hg)	Wind velocity (mph) $V_{9''}$	$V_{6''}$	$V_{30'}$	$V_{13'}$	Evaporation (pan) (in./mo)	Evaporation estimates (in./mo) Pan	Lake Hefner	Rohwer	Meyer
Apr.	48.0	0.34	0.22	0.34	64	29.96	4.0	1.7	24.8	20.0	4.60	3.22	5.37	1.62	4.75
May	58.5	0.52	0.31	0.50	62	29.91	3.1	1.3	19.4	15.5	5.76	4.03	7.59	2.65	5.82
June	71.9	0.72	0.48	0.78	61	29.89	2.0	0.9	12.4	10.0	7.76	5.43	5.38	2.75	6.12
July	76.3	0.86	0.53	0.91	58	30.00	1.7	0.7	10.5	8.5	9.36	6.55	6.45	4.45	6.58
Aug.	71.3	0.72	0.58	0.77	75	29.99	1.3	0.6	8.1	6.5	5.58	3.91	2.10	2.42	2.45
Sept.	64.9	0.62	0.40	0.62	64	30.08	1.4	0.6	8.7	7.0	4.98	3.49	3.45	2.33	3.22
Oct.	57.2	0.50	0.28	0.47	59	30.01	1.9	0.8	11.8	9.5	3.80	2.66	4.83	2.55	3.70
Totals												29.29	35.17	18.77	32.64

TABLE 10.3. Water budget for Lake MacBride

Month	Stream inflow (acre-ft/mo)	Precipitation (acre-ft/mo)	Evaporation loss (acre-ft/mo)	Seepage loss (acre-ft/mo)	Total inflow (acre-ft)	Total outflow (acre-ft)	Net storage (acre-ft)	Elevation change (ft)	Surface area (acres)	Elevation at end of month (ft)
March									960	712.0
Apr.	652	243	430	120	895	550	+345 goes over spillway	+0.4	1,328	712.4
May	242	169	610	124	411	734	−323	−0.3	1,052	712.1
June	235	221	430	120	456	550	−94	−0.1	960	712.0
July	7	223	512	124	230	636	−406	−0.4	592	711.6
Aug.	35	329	168	124	364	292	+72	+0.1	684	711.7
Sept.	57	183	276	120	240	396	−156	−0.2	500	711.5
Oct.	7	175	387	124	182	511	−329	−0.4	132	711.1
Total	1,235	1,543	2,813	856	2,778	3,669	−891	−0.9		

The subscript 0 refers to a known vertical location (9 inches for this problem).

For this particular study the Lake Hefner equation, Eq. (10.8), gives the largest values of evaporation. It should be noted that in some months a variation of more than 100% exists in the predicted values. A large value of evaporation will be used here because it provides for a conservative decision.

A seepage analysis of the earthen dam indicates that 2 cfs can be expected as seepage loss. When the data in Tables 10.1 and 10.2 are used, the water budget is accomplished as shown in Table 10.3. Note that this is actually a simulation of the reservoir's behavior but for only a portion of one year. In the final analysis the enlargement of the dam may not be feasible from a physical standpoint. If possible a longer period of record should be studied both by simulation and by the stochastic methods of Chapter 8. If the project proves physically feasible its economic feasibility should be studied by the methods of Chapter 3.

For further discussion of evaporation the reader is referred to the work of Morton (18), Penman (19), and Lane (20).

PROBLEMS

10.1. Death Valley in California has experienced 155 inches of evaporation during at least one year. Compare this to a reasonable evaporation figure for your hometown.

10.2. Compare average annual evaporation with average annual precipitation for (a) your hometown, (b) Houston, Tex., (c) Seattle, Wash., and (d) Salt Lake City, Utah.

10.3. The All-American canal in southern California has a mean water surface width of 200 feet. Assuming water temperature of 75° F, air temperature of 95° F, and relative humidity of 6%, estimate the monthly cost of evaporated water. Water delivered through the canal is worth $5.50/acre-foot.

10.4. Using U.S. Weather Bureau climatological data and U.S. Geological Survey water supply data, construct a water budget for a reservoir location in your area.

10.5. During August of a particular year evaporation from a class A land pan at a lake location indicated a total evaporation of 18 inches. If the reservoir surface area was 6000 acres on August 1 and decreases at the rate of 500 acres/foot of depth, calculate the evaporation loss from the lake.

10.6. Using climatological data as given in the example problem for this chapter, calculate evaporation estimates using each of the Dalton type of equations.

10.7. A lake in southern Missouri that is rarely frozen over has a water surface area of 4140 acres when full. Climatological data for an average year is as follows:

Month	Average air temperature (°F)	Pan evaporation (in.)	Wind speed, V_{13} (mph)	Relative humidity (%)	Barometric pressure (in. Hg)
January	38	1.5	2.5	53	30.10
February	40	2.6	3.0	55	30.06
March	42	3.0	4.0	60	30.01
April	50	3.8	3.8	64	29.96
May	61	4.4	3.4	67	29.92
June	66	6.5	2.0	70	29.90
July	75	7.9	2.0	71	29.91
August	79	8.7	2.1	70	29.98
September	60	6.7	1.8	63	30.01
October	49	4.8	1.8	56	30.14
November	41	3.5	2.2	54	30.08
December	38	1.6	2.5	54	30.01

Estimate the difference between annual precipitation and evaporation for this lake. Based on this year, what would be the required annual inflow to the lake just to maintain a full reservoir without downstream releases?

REFERENCES

1. Kohler, M. A.; Nordenson, T. J.; and Baker, D. R. Evaporation maps for the United States. U.S. Weather Bureau Tech. Paper 37, Washington, D.C., 1959.
2. List, R. J. *Smithsonian Meteorological Tables*, 6th ed. rev. The Smithsonian Institution, Washington, D.C., 1951.
3. Raphael, J. M. Prediction of temperature in rivers and reservoirs. *ASCE J. Power Div.* No. PO2, Paper 3200, July 1962.
4. Evaporation from water surfaces. A symposium. Trans. ASCE, Vol. 99, 1934.
5. U.S. Weather Bureau. Climatological data. National summary, monthly, 1950–.
6. Dalton, J. Experimental essays on the constitution of mixed gases; on the force of steam or vapor from water and other liquids in different temperatures, both in a Torricellian vacuum and in air; on evaporation; and on expansion of gases by heat. Manchester Lit. Phil. Soc. Mem. Proc., Vol. 5, 1802.
7. Fitzgerald, D. Evaporation. Trans. ASCE, Vol. 15, 1886.
8. Russell, T. Depth of evaporation in the United States. *Monthly Weather Rev.*, Vol. 16, 1888.

9. Horton, R. E. A new evaporation formula developed. *Eng. News Record,* Vol. 78, No. 4, Apr. 26, 1917.

10. Rohwer, C. Evaporation from free water surface. USDA Tech. Bull. 271, Washington, D.C., 1931.

11. Meyer, A. F. Computing runoff from rainfall and other physical data. Trans. ASCE, Vol. 79, 1915.

12. Harbeck, G. E.; and Anderson, E. R. Water-loss investigations: Vol. 1— Lake Hefner studies technical report. U.S. Geological Survey Paper 269, Washington, D.C., 1954.

13. Kohler, M. A.; Nordenson, T. J.; and Fox, W. E. Evaporation from pans and lakes. U.S. Weather Bureau Research Paper 38, Washington, D.C., May 1955.

14. Anderson, E. R. Energy-budget studies from water loss investigations: Vol. 1—Lake Hefner studies technical report. U.S. Geological Survey Circ. 229, Washington, D.C., 1952.

15. Eagleson, P. S. *Dynamic Hydrology.* McGraw-Hill, 1970.

16. Bowen, I. S. The ratio of heat losses by conduction and by evaporation from any water surface. *Phys. Rev.,* Ser. 2, Vol. 27, June 1926.

17. Harbeck, G. E.; and Meyers, J. S. Present day evaporation measurement techniques. *ASCE J. Hydraul. Div.,* Vol. 96, July 1970.

18. Morton, F. I. Potential evaporation and river basin evaporation. *ASCE J. Hydraul. Div.,* Vol. 91, Nov. 1965.

19. Penman, H. L. Natural evaporation from open water, bare soil, and grass. Proc. Roy. Soc. London, Ser. A, Vol. 193, No. 1032, 1948.

20. Lane, R. K. Estimating evaporation from insolation. *ASCE J. Hydraul. Div.,* Vol. 90, Sept. 1964.

APPENDICES

APPENDIX A

GENERAL STATISTICAL DATA

TABLE A.1. Miscellaneous conversion factors

Multiply	By	To obtain
cfs-days	1.983	acre-feet
cfs-days	0.03719	inches depth on 1 square mile
cfs-days/square mile	0.03719	inches depth
cfs-hours	0.08264	acre-feet
cfs-hours/square mile	0.001550	inches depth
cfs	1.983	acre-feet/day
cfs	724.0	acre-feet/year (365 days)
cfs	448.8	U.S. gallons/minute
cfs	0.6463	million U.S. gallons/day
csm (cfs/square mile)	0.03719	inches depth/day
csm	13.57	inches depth/year (365 days)
inches/hour	645.3	csm
inches/hour	1.008	cfs/acre
inches depth	53.33	acre-feet/square mile
inches depth on 1 square mile	53.33	acre-feet
acre-feet	0.5042	cfs-days
acre-feet	12.10	cfs-hours
acre-feet	0.01875	inches depth on 1 square mile
acre-feet	0.3258	million U.S. gallons
acre-feet/day	0.5042	cfs
acre-feet/square mile	0.01875	inches depth
U.S. gallons/minute	0.002228	cfs
million U.S. gallons/day	1.547	cfs
million U.S. gallons/day	3.069	acre-feet
feet/second	0.6818	miles/hour
centimeters	0.3937	inches
hectares	2.471	acres
liters	0.2642	U.S. gallons
kilograms	2.205	pounds
cubic feet	7.480	U.S. gallons
imperial gallons	1.200	U.S. gallons

Source: U.S. Soil Conservation Service. Hydrology. *In* National Engineering Handbook, Sect. 4, Part 1, 1964.

TABLE A.2. Conversion factors—British to metric units of measurement

The following conversion factors adopted by the Bureau of Reclamation are those published by the American Society for Testing and Materials (ASTM Metric Practice Guide, E 380-68) except that additional factors (*) commonly used in the Bureau have been added. Further discussion of definitions of quantities and units is given in the ASTM Metric Practice Guide.

The metric units and conversion factors adopted by the ASTM are based on the "International System of Units" (designated SI for Système International d'Unités), fixed by the International Committee for Weights and Measures; this system is also known as the Giorgi or MKSA (meter-kilogram [mass]-second-ampere) system. This system has been adopted by the International Organization for Standardization in ISO Recommendation R-31.

The metric technical unit of force is the kilogram-force; this is the force which, when applied to a body having a mass of 1 kg, gives it an acceleration of 9.80665 m/sec/sec, the standard acceleration of free fall toward the earth's center for sea level at 45° latitude. The metric unit of force in SI units is the newton (N), which is defined as that force which, when applied to a body having a mass of 1 kg, gives it an acceleration of 1 m/sec/sec. These units must be distinguished from the (inconstant) local weight of a body having a mass of 1 kg; i.e., the weight of a body is that force with which a body is attracted to the earth and is equal to the mass of a body multiplied by the acceleration due to gravity. However, because it is general practice to use "pound" rather than the technically correct term "pound-force," the term "kilogram" (or derived mass unit) has been used in this guide instead of "kilogram-force" in expressing the conversion factors for forces. The newton unit of force will find increasing use, and is essential in SI units.

Where approximate or nominal British units are used to express a value or range of values, the converted metric units in parentheses are also approximate or nominal. Where precise British units are used, the converted metric units are expressed as equally significant values.

Multiply	By	To obtain
Length		
Mil	25.4 (exactly)	micron
Inches	25.4 (exactly)	millimeters
Inches	2.54 (exactly)*	centimeters
Feet	30.48 (exactly)	centimeters
Feet	0.3048 (exactly)*	meters
Feet	0.0003048 (exactly)*	kilometers
Yards	0.9144 (exactly)	meters
Miles (statute)	1,609.344 (exactly)*	meters
Miles	1.609344 (exactly)	kilometers
Area		
Square inches	6.4516 (exactly)	square centimeters
Square feet	*929.03	square centimeters
Square feet	0.092903	square meters
Square yards	0.836127	square meters
Acres	*0.40469	hectares
Acres	*4,046.9	square meters
Acres	*0.0040469	square kilometers
Square miles	2.58999	square kilometers
Volume		
Cubic inches	16.3871	cubic centimeters
Cubic feet	0.0283168	cubic meters
Cubic yards	0.764555	cubic meters

TABLE A.2. (continued)

Multiply	By	To obtain
Capacity		
Fluid ounces (U.S.)	29.5737	cubic centimeters
Fluid ounces (U.S.)	29.5729	milliliters
Liquid pints (U.S.)	0.473179	cubic decimeters
Liquid pints (U.S.)	0.473166	liters
Quarts (U.S.)	*946.358	cubic centimeters
Quarts (U.S.)	*0.946331	liters
Gallons (U.S.)	*3,785.43	cubic centimeters
Gallons (U.S.)	3.78543	cubic decimeters
Gallons (U.S.)	3.78533	liters
Gallons (U.S.)	*0.00378543	cubic meters
Gallons (U.K.)	4.54609	cubic decimeters
Gallons (U.K.)	4.54596	liters
Cubic feet	28.3160	liters
Cubic yards	*764.55	liters
Acre-feet	*1,233.5	cubic meters
Acre-feet	*1,233,500	liters
Mass		
Grains (1/7,000 lb)	64.79891 (exactly)	milligrams
Troy ounces (480 grains)	31.1035	grams
Ounces (avdp)	28.3495	grams
Pounds (avdp)	0.45359237 (exactly)	kilograms
Short tons (2,000 lb)	907.185	kilograms
	0.907185	metric tons
Long tons (2,240 lb)	1,016.05	kilograms
Force/Area		
Pounds/square inch	0.070307	kilograms/square centimeter
	0.689476	newtons/square centimeter
Pounds/square foot	4.88243	kilograms/square meter
	47.8803	newtons/square meter
Mass/Volume (Density)		
Ounces/cubic inch	1.72999	grams/cubic centimeter
Pounds/cubic foot	16.0185	kilograms/cubic meter
	0.0160185	grams/cubic centimeter
Tons (long)/cubic yard	1.32894	grams/cubic centimeter
Mass/Capacity		
Ounces/gallon (U.S.)	7.4893	grams/liter
Ounces/gallon (U.K.)	6.2362	grams/liter
Pounds/gallon (U.S.)	119.829	grams/liter
Pounds/gallon (U.K.)	99.779	grams/liter
Bending Moment or Torque		
Inch-pounds	0.011521	meter-kilograms
	1.12985×10^6	centimeter-dynes
Foot-pounds	0.138255	meter-kilograms
	1.35582×10^7	centimeter-dynes
Foot-pounds/inch	5.4431	centimeter-kilograms/centimeter
Ounce-inches	72.008	gram-centimeters

TABLE A.2. (continued)

Multiply	By	To obtain
	Velocity	
Feet/second	30.48 (exactly)	centimeters/second
	0.3048 (exactly)*	meters/second
Feet/year	0.965873×10^{-6}*	centimeters/second
Miles/hour	1.609344 (exactly)	kilometers/hour
	0.44704 (exactly)	meters/second
	*Acceleration**	
Feet/second²	0.3048*	meters/second²
	Flow	
Cubic feet/second (second-feet)	0.028317*	cubic meters/second
Cubic feet/minute	0.4719	liters/second
Gallons (U.S.)/minute	0.06309	liters/second
	*Force**	
Pounds	0.453592*	kilograms
	4.4482*	newtons
	4.4482×10^{-5}*	dynes
	*Work and Energy**	
British thermal units (Btu)	0.252*	kilogram calories
	1,055.06	joules
Btu/pound	2.326 (exactly)	joules/gram
Foot-pounds	1.35582*	joules
	Power	
Horsepower	745.700	watts
Btu/hour	0.293071	watts
Foot-pounds/second	1.35582	watts
	Heat Transfer	
Btu in./hr ft² °F (k, thermal conductivity)	1.442	milliwatts/cm °C
	0.1240	kg cal/hr m °C
Btu ft/hr ft² °F	1.4880*	kg cal m/hr in.² °C
Btu/hr ft² °F (C, thermal conductance)	0.568	milliwatts/cm² °C
	4.882	kg cal/hr m² °C
°F hr ft²/Btu (R, thermal resistance)	1.761	°C cm²/milliwatt
Btu/lb °F (c, heat capacity)	4.1868	J/g °C
Btu/lb °F	1.000*	cal/gram °C
Ft²/hr (thermal diffusivity)	0.2581	cm²/sec
	0.09290*	m²/hr
	Water Vapor Transmission	
Grains/hr ft² (water vapor transmission)	16.7	grams/24 hr m²
Perms (permeance)	0.659	metric perms
Perm-inches (permeability)	1.67	metric perm-centimeters

TABLE A.2.   (continued)

Multiply	By	To obtain
Other Quantities and Units		
Cubic feet/square foot/day (seepage)	304.8*	liters/square meter/day
Pound-seconds/square foot (viscosity)	4.8824*	kilogram second/square meter
Square feet/second (viscosity)	0.092903*	square meters/second
Fahrenheit degrees (change)*	5/9 (exactly)	Celsius or Kelvin degrees (change)*
Volts per mil	0.03937	kilovolts/millimeter
Lumens/square foot (foot-candles)	10.764	lumens/square meter
Ohm-circular mils/foot	0.001662	ohm-square millimeters/meter
Millicuries/cubic foot	35.3147*	millicuries/cubic meter
Milliamps/square foot	10.7639*	milliamps/square meter
Gallons/square yard	4.527219*	liters/square meter
Pounds/inch	0.17858*	kilograms/centimeter

TABLE A.3. Variation of dewpoint with temperature and wet-bulb depression and of saturation vapor pressure with temperature (pressure = 30 in.)

Air temperature (°F)	Saturation vapor pressure Milli-bars	Saturation vapor pressure In. Hg	Wet-bulb depression (°F) 1	2	3	4	6	8	10	12	14	16	18	20	25	30
0	1.29	0.038	-7	-20												
5	1.66	0.049	-1	-9	-24											
10	2.13	0.063	5	-2	-10	-27										
15	2.74	0.081	11	6	0	-9										
20	3.49	0.103	16	12	8	2	-21									
25	4.40	0.130	22	19	15	10	-3	-15								
30	5.55	0.164	27	25	21	18	8	-7								
35	6.87	0.203	33	30	28	25	17	7	-11							
40	8.36	0.247	38	35	33	30	25	18	7							
45	10.09	0.298	43	41	38	36	31	25	18	-14						
50	12.19	0.360	48	46	44	42	37	32	26	7	-14					
55	14.63	0.432	53	51	50	48	43	38	33	18	8	-13				
60	17.51	0.517	58	57	55	53	49	45	40	27	20	9	-12			
65	20.86	0.616	63	62	60	59	55	51	47	35	29	21	11	-8		
70	24.79	0.732	69	67	65	64	61	57	53	42	37	31	24	14	-11	
75	29.32	0.866	74	72	71	69	66	63	59	49	44	39	33	26	15	
80	34.61	1.022	79	77	76	74	68	65	62	55	51	47	42	36	28	
85	40.67	1.201	84	82	81	80	77	74	71	68	64	61	57	52	39	19
90	47.68	1.408	89	87	86	85	82	79	76	73	70	67	63	59	48	32
95	55.71	1.645	94	93	91	90	87	85	82	79	76	73	70	66	56	43
100	64.88	1.916	99	98	96	95	93	90	87	85	82	79	76	72	63	52

TABLE A.4. Variation of pressure, temperature, and boiling point with elevation (U.S. standard atmosphere)

Elevation (ft msl)	Pressure			Air temperature (°F)	Boiling point (°F)
	In. Hg	Millibars	Feet of water		
—1,000	31.02	1,050.5	35.12	62.6	213.8
0	29.92	1,013.2	33.87	59.0	212.0
1,000	28.86	977.3	32.67	55.4	210.2
2,000	27.82	942.1	31.50	51.8	208.4
3,000	26.81	907.9	30.35	48.4	206.5
4,000	25.84	875.0	29.25	44.8	204.7
5,000	24.89	842.9	28.18	41.2	202.9
6,000	23.98	812.1	27.15	37.6	201.1
7,000	23.09	781.9	26.14	34.0	199.2
8,000	22.22	752.5	25.16	30.6	197.4
9,000	21.38	724.0	24.20	27.0	195.6
10,000	20.58	696.9	23.30	23.4	193.7
11,000	19.79	670.2	22.40	19.8	191.9
12,000	19.03	644.4	21.54	16.2	190.1

Note: The data of this table are based on average conditions.

TABLE A.5. Variation of relative humidity (%) with temperature and wet-bulb depression (pressure = 30 in.)

Air temperature (°F)	Wet-bulb depression (°F)													
	1	2	3	4	6	8	10	12	14	16	18	20	25	30
0	67	33	1											
5	73	46	20											
10	78	56	34	13										
15	82	64	46	29										
20	85	70	55	40	12									
25	87	74	62	49	25	1								
30	89	78	67	56	36	16								
35	91	81	72	63	45	27	10							
40	92	83	75	68	52	37	22	7						
45	93	86	78	71	57	44	31	18	6					
50	93	87	80	74	61	49	38	27	16	5				
55	94	88	82	76	65	54	43	33	23	14	5			
60	94	89	83	78	68	58	48	39	30	21	13	5		
65	95	90	85	80	70	61	52	44	35	27	20	12		
70	95	90	86	81	72	64	55	48	40	33	25	19	3	
75	96	91	86	82	74	66	58	51	44	37	30	24	9	
80	96	91	87	83	75	68	61	54	47	41	35	29	15	3
85	96	92	88	84	76	70	63	56	50	44	38	32	20	8
90	96	92	89	85	78	71	65	58	52	47	41	36	24	13
95	96	93	89	86	79	72	66	60	54	49	44	38	27	17
100	96	93	89	86	80	73	68	62	56	51	46	41	30	21

TABLE A.6. Properties of water

Temperature (°F)	Specific gravity	Unit weight (lb/cu ft)	Heat of vaporization (Btu/lb)	Viscosity Absolute (lb-sec/sq ft)	Viscosity Kinematic (sq ft/sec)	Vapor pressure Millibars	Vapor pressure Psi
32	0.99987	62.416	1,073	0.374×10^{-4}	1.93×10^{-5}	6.11	0.09
40	0.99999	62.423	1,063	0.323	1.67	8.36	0.12
50	0.99975	62.408	1,059	0.273	1.41	12.19	0.18
60	0.99907	62.366	1,054	0.235	1.21	17.51	0.26
70	0.99802	62.300	1,049	0.205	1.06	24.79	0.36
80	0.99669	62.217	1,044	0.180	0.929	34.61	0.51
90	0.99510	62.118	1,039	0.160	0.828	47.68	0.70
100	0.99318	61.998	1,033	0.143	0.741	64.88	0.95
120	0.98870	61.719	1,021	0.117	0.610	...	1.69
140	0.98338	61.386	1,010	0.0979	0.513	...	2.89
160	0.97729	61.006	999	0.0835	0.440	...	4.74
180	0.97056	60.586	988	0.0726	0.385	...	7.51
200	0.96333	60.135	977	0.0637	0.341	...	11.52
212	0.95865	59.843	970	0.0593	0.319	...	14.70

TABLE A.7.   Uniformly distributed random numbers

53 74 23 99 67	61 32 28 69 84	94 62 67 86 24	98 33 41 19 95	47 53 53 38 09
63 38 06 86 54	99 00 65 26 94	02 82 90 23 07	79 62 67 80 60	75 91 12 81 19
35 30 58 21 46	06 72 17 10 94	25 21 31 75 96	49 28 24 00 49	55 65 79 78 07
63 43 36 82 69	65 51 18 37 88	61 38 44 12 45	32 92 85 88 65	54 34 81 85 35
98 25 37 55 26	01 91 82 81 46	74 71 12 94 97	24 02 71 37 07	03 92 18 66 75
02 63 21 17 69	71 50 80 89 56	38 15 70 11 48	43 40 45 86 98	00 83 26 91 03
64 55 22 21 82	48 22 28 06 00	61 54 13 43 91	82 78 12 23 29	06 66 24 12 27
85 07 26 13 89	01 10 07 82 04	59 63 69 36 03	69 11 15 83 80	13 29 54 19 28
58 54 16 24 15	51 54 44 82 00	62 61 65 04 69	38 18 65 18 97	85 72 13 49 21
34 85 27 84 87	61 48 64 56 26	90 18 48 13 26	37 70 15 42 57	65 65 80 39 07
03 92 18 27 46	57 99 16 96 56	30 33 72 85 22	84 64 38 56 98	99 01 30 98 64
62 95 30 27 59	37 75 41 66 48	86 97 80 61 45	23 53 04 01 63	45 76 08 64 27
08 45 93 15 22	60 21 75 46 91	98 77 27 85 42	28 88 61 08 84	69 62 03 42 73
07 08 55 18 40	45 44 75 13 90	24 94 96 61 02	57 55 66 83 15	73 42 37 11 61
01 85 89 95 66	51 10 19 34 88	15 84 97 19 75	12 76 39 43 78	64 63 91 08 25
72 84 71 14 35	19 11 58 49 26	50 11 17 17 76	86 31 57 20 18	95 60 78 46 75
88 78 28 16 84	13 52 53 94 53	75 45 69 30 96	73 89 65 70 31	99 17 43 48 76
45 17 75 65 57	28 40 19 72 12	25 12 74 75 67	60 40 60 81 19	24 62 01 61 16
96 76 28 12 54	22 01 11 94 25	71 96 16 16 88	68 64 36 74 45	19 59 50 88 92
43 31 67 72 30	24 02 94 08 63	38 32 36 66 02	69 36 38 25 39	48 03 45 15 22
50 44 66 44 21	66 06 58 05 62	68 15 54 35 02	42 35 48 96 32	14 52 41 52 48
22 66 22 15 86	26 63 75 41 99	58 42 36 72 24	58 37 52 18 51	03 37 18 39 11
96 24 40 14 51	23 22 30 88 57	95 67 47 29 83	94 69 40 06 07	18 16 36 78 86
31 73 91 61 19	60 20 72 93 48	98 57 07 23 69	65 95 39 69 58	56 80 30 19 44
78 60 73 99 84	43 89 94 36 45	56 69 47 07 41	90 22 91 07 12	78 35 34 08 72
84 37 90 61 56	70 10 23 98 05	85 11 34 76 60	76 48 45 34 60	01 64 18 39 96
36 67 10 08 23	98 93 35 08 86	99 29 76 29 81	33 34 91 58 93	63 14 52 32 52
07 28 59 07 48	89 64 58 89 75	83 85 62 27 89	30 14 78 56 27	86 63 59 80 02
10 15 83 87 60	79 24 31 66 56	21 48 24 06 93	91 98 94 05 49	01 47 59 38 00
55 19 68 97 65	03 73 52 16 56	00 53 55 90 27	33 42 29 38 87	22 13 88 83 34
53 81 29 13 39	35 01 20 71 34	62 33 74 82 14	53 73 19 09 03	56 54 29 56 93
51 86 32 68 92	33 98 74 66 99	40 14 71 94 58	45 94 19 38 81	14 44 99 81 07
35 91 70 29 13	80 03 54 07 27	96 94 78 32 66	50 95 52 74 33	13 80 55 62 54
37 71 67 95 13	20 02 44 95 94	64 85 04 05 72	01 32 90 76 14	53 89 74 60 41
93 66 13 83 27	92 79 64 64 72	28 54 96 53 84	48 14 52 98 94	56 07 93 89 30
02 96 08 45 65	13 05 00 41 84	93 07 54 72 59	21 45 57 09 77	19 48 56 27 44
49 83 43 48 35	82 88 33 69 96	72 36 04 19 76	47 45 15 18 60	82 11 08 95 97
84 60 71 62 46	40 80 81 30 37	34 39 23 05 38	25 15 35 71 30	88 12 57 21 77
18 17 30 88 71	44 91 14 88 47	89 23 30 63 15	56 34 20 47 89	99 82 93 24 98
79 69 10 61 78	71 32 76 95 62	87 00 22 58 40	92 54 01 75 25	43 11 71 99 31
75 93 36 57 83	56 20 14 82 11	74 21 97 90 65	96 42 68 63 86	74 54 13 26 94
38 30 92 29 03	06 28 81 39 38	62 25 06 84 63	61 29 08 93 67	04 32 92 08 09
51 29 50 10 34	31 57 75 95 80	51 97 02 74 77	76 15 48 49 44	18 55 63 77 09
21 31 38 86 24	37 79 81 53 74	73 24 16 10 33	52 83 90 94 76	70 47 14 54 36
29 01 23 87 88	58 02 39 37 67	42 10 14 20 92	16 55 23 42 45	54 96 09 11 06
95 33 95 22 00	18 74 72 00 18	38 79 58 69 32	81 76 80 26 92	82 80 84 25 39
90 84 60 79 80	24 36 59 87 38	82 07 53 89 35	96 35 23 79 18	05 98 90 07 35
46 40 62 98 82	54 97 20 56 95	15 74 80 08 32	16 46 70 50 80	67 72 16 42 79
20 31 89 03 43	38 46 82 68 72	32 14 82 99 70	80 60 47 18 97	63 49 30 21 30
71 59 73 05 50	08 22 23 71 77	91 01 93 20 49	82 96 59 26 94	66 39 67 98 60

Source: L. R. Beard. *Statistical Methods in Hydrology.* U.S. Army, Corps of Engineers, 1962.

TABLE A.8. **Common logarithms**

	0	1	2	3	4	5	6	7	8	9
10	000	004	009	013	017	021	025	029	033	037
11	041	045	049	053	057	061	064	068	072	076
12	079	083	086	090	093	097	100	104	107	111
13	114	117	121	124	127	130	134	137	140	143
14	146	149	152	155	158	161	164	167	170	173
15	176	179	182	185	188	190	193	196	199	201
16	204	207	210	212	215	217	220	223	225	228
17	230	233	236	238	241	243	246	248	250	253
18	255	258	260	262	265	267	270	272	274	276
19	279	281	283	286	288	290	292	294	297	299
20	301	303	305	308	310	312	314	316	318	320
21	322	324	326	328	330	332	334	336	338	340
22	342	344	346	348	350	352	354	356	358	360
23	362	364	366	367	369	371	373	375	377	378
24	380	382	384	386	387	389	391	393	394	396
25	398	400	401	403	405	407	408	410	412	413
26	415	417	418	420	422	423	425	427	428	430
27	431	433	435	436	438	439	441	442	444	446
28	447	449	450	452	453	455	456	458	459	461
29	462	464	465	467	468	470	471	473	474	476
30	477	479	480	481	483	484	486	487	489	490
31	491	493	494	496	497	498	500	501	502	504
32	505	507	508	509	511	512	513	515	516	517
33	519	520	521	522	524	525	526	528	529	530
34	531	533	534	535	537	538	539	540	542	543
35	544	545	547	548	549	550	551	553	554	555
36	556	558	559	560	561	562	563	565	566	567
37	568	569	571	572	573	574	575	576	577	579
38	580	581	582	583	584	585	587	588	589	590
39	591	592	593	594	596	597	598	599	600	601
40	602	603	604	605	606	607	609	610	611	612
41	613	614	615	616	617	618	619	620	621	622
42	623	624	625	626	627	628	629	630	631	632
43	633	634	635	636	637	638	639	640	641	642
44	643	644	645	646	647	648	649	650	651	652
45	653	654	655	656	657	658	659	660	661	662
46	663	664	665	666	667	667	668	669	670	671
47	672	673	674	675	676	677	678	679	679	680
48	681	682	683	684	685	686	687	688	688	689
49	690	691	692	693	694	695	695	696	697	698
50	699	700	701	702	702	703	704	705	706	707
51	708	708	709	710	711	712	713	713	714	715
52	716	717	718	719	719	720	721	722	723	723
53	724	725	726	727	728	728	729	730	731	732
54	732	733	734	735	736	736	737	738	739	740

TABLE A.8.   (continued)

	0	1	2	3	4	5	6	7	8	9
55	740	741	742	743	744	744	745	746	747	747
56	748	749	750	751	751	752	753	754	754	755
57	756	757	757	758	759	760	760	761	762	763
58	763	764	765	766	766	767	768	769	769	770
59	771	772	772	773	774	775	775	776	777	777
60	778	779	780	780	781	782	782	783	784	785
61	785	786	787	787	788	789	790	790	791	792
62	792	793	794	794	795	796	797	797	798	799
63	799	800	801	801	802	803	803	804	805	806
64	806	807	808	808	809	810	810	811	812	812
65	813	814	814	815	816	816	817	818	818	819
66	820	820	821	822	822	823	823	824	825	825
67	826	827	827	828	829	829	830	831	831	832
68	833	833	834	834	835	836	836	837	838	838
69	839	839	840	841	841	842	843	843	844	844
70	845	846	846	847	848	848	849	849	850	851
71	851	852	852	853	854	854	855	856	856	857
72	857	858	859	859	860	860	861	862	862	863
73	863	864	865	865	866	866	867	867	868	869
74	869	870	870	871	872	872	873	873	874	875
75	875	876	876	877	877	878	879	879	880	880
76	881	881	882	883	883	884	884	885	885	886
77	886	887	888	888	889	889	890	890	891	892
78	892	893	893	894	894	895	895	896	897	897
79	898	898	899	899	900	900	901	902	902	903
80	903	904	904	905	905	906	906	907	907	908
81	908	909	910	910	911	911	912	912	913	913
82	914	914	915	915	916	916	917	918	918	919
83	919	920	920	921	921	922	922	923	923	924
84	924	925	925	926	926	927	927	928	928	929
85	929	930	930	931	931	932	932	933	933	934
86	935	935	936	936	937	937	938	938	939	939
87	940	940	941	941	942	942	943	943	943	944
88	944	945	945	946	946	947	947	948	948	949
89	949	950	950	951	951	952	952	953	953	954
90	954	955	955	956	956	957	957	958	958	959
91	959	960	960	960	961	961	962	962	963	963
92	964	964	965	965	966	966	967	967	968	968
93	968	969	969	970	970	971	971	972	972	973
94	973	974	974	975	975	975	976	976	977	977
95	978	978	979	979	980	980	980	981	981	982
96	982	983	983	984	984	985	985	985	986	986
97	987	987	988	988	989	989	989	990	990	991
98	991	992	992	993	993	993	994	994	995	995
99	996	996	997	997	997	998	998	999	999	999

BLUE RIVER STREAMFLOW DATA

BLUE RIVER

Location—Lat. 38°57'25", long. 94°33'32", in SE ¼ NE ¼ sec. 28, T.48N, R.33W, on downstream side of right pier of bridge on County Highway W, 0.4 mile downstream from Indian Creek and 1.7 miles southeast of Kansas City.

Drainage area—188 square miles. *Slope*—12.4 feet/mile.

Gage—Nonrecording prior to July 1, 1939; recording gage thereafter. Datum of gage is 753.73 feet above mean sea level (levels by Corps of Engineers).

Stage-discharge relation—Defined by current-meter measurements.

Bank-full stage—14 feet.

Historical data—Maximum stage known prior to 1961, about 39 feet on November 17, 1928, occurred before construction of present bridge and major changes in channel at gage site.

Remarks—Base for partial-duration series, 5800 cfs.

FIG. B.1.   Basin of the Blue River.

TABLE B.1. Discharge from the Blue River basin

Water year	Date	Peak discharge (cfs)	Mean annual discharge (cfs)
1939	June 25, 1939	8,140	
1940	Apr. 27, 1940	5,990	77
	May 18, 1940	6,250	
	June 23, 1940	7,000	
1941	Apr. 4, 1941	6,460	125
1942	Oct. 31, 1941	6,730	189
	June 19, 1942	7,280	
	July 25, 1942	7,890	
1943	June 10, 1943	5,650	133
1944	Apr. 23, 1944	26,400	165
	May 21, 1944	7,010	
1945	Mar. 24, 1945	6,000	270
	Apr. 16, 1945	11,100	
	May 16, 1945	8,460	
	June 30, 1945	8,740	
1946	May 10, 1946	7,890	76
1947	Mar. 13, 1947	7,780	220
	Apr. 3, 1947	7,620	
	Apr. 5, 1947	12,100	
	Apr. 10, 1947	7,120	
	June 21, 1947	8,120	
	June 23, 1947	14,100	
1948	Mar. 19, 1948	7,970	137
	July 22, 1948	7,970	
	July 26, 1948	9,540	
1949	May 21, 1949	7,180	148
	June 6, 1949	8,800	
	June 7, 1949	6,200	
1950	Oct. 21, 1949	16,400	162
	July 12, 1950	6,200	
	Aug. 27, 1950	7,180	
1951	June 26, 1951	7,350	290
	June 29, 1951	6,580	
	July 6, 1951	7,740	
	July 11, 1951	31,100	
	Sept. 4, 1951	6,200	
	Sept. 9, 1951	6,800	
1952	Mar. 10, 1952	8,380	133
1953	Apr. 30, 1953	1,760	22
1954	Aug. 2, 1954	4,650	23
1955	Oct. 20, 1954	6,360	91
	May 28, 1955	8,560	
1956	Oct. 5, 1955	1,270	13
1957	May 16, 1957	6,710	56
	June 30, 1957	14,300	

TABLE B.1. (continued)

Water year	Date	Peak discharge (cfs)	Mean annual discharge (cfs)
1958	July 17, 1958	9,180	231
	July 20, 1958	6,160	
	July 25, 1958	6,640	
	July 31, 1958	21,700	
	Aug. 16, 1958	7,900	
1959	Apr. 27, 1959	5,120	65
1960	Apr. 16, 1960	7,980	98
	Apr. 30, 1960	7,900	
1961	May 6, 1961	8,200	285
	July 6, 1961	7,780	
	Sept. 13, 1961	41,000	
	Sept. 24, 1961	7,430	
1962	Nov. 2, 1961	9,140	194
	Nov. 16, 1961	7,090	
1963	July 13, 1963	4,390	64
1964	May 26, 1964	7,090	113
	May 28, 1964	8,130	
1965	June 5, 1965	12,100	178
	Sept. 4, 1965	9,050	
1966	July 13, 1966	5,520	80
1967	June 21, 1967	14,600	186
	June 24, 1967	10,100	
1968	Oct. 5, 1967	4,880	138
1969	June 21, 1969	9,160	184
	June 27, 1969	14,200	
1970	Oct. 12, 1969	8,760	217
	Apr. 18, 1970	6,630	
	June 12, 1970	8,280	
	Sept. 16, 1970	6,630	
	Sept. 22, 1970	15,400	
1971	Jan. 3, 1970	7,660	117

TABLE B.2. Monthly and yearly mean total flow from the Blue River basin (cubic feet/second)

Water year	Oct.	Nov.	Dec.	Jan.	Feb.	Mar.	Apr.	May	June	July	Aug.	Sept.	The year
1939	...	...	...	0	...	...	...	84.8	261	4.90	82.0	0.05	...
1940	0	0	0	0	2.66	32.1	161	297	238	19.3	158	17.4	77.2
1941	0.66	41.9	49.7	445	197	664	390	59.2	166	52.1	2.94	41.6	125
1942	463	189	101	47.0	113	136	304	203	312	184	127	78.1	189
1943	103	149	250	95.2	59.5	49.0	42.7	334	372	55.0	77.6	1.84	133
1944	3.76	2.31	2.85	11.3	17.7	203	1,279	290	60.7	10.5	96.7	11.9	165
1945	30.9	24.7	251	47.2	68.5	499	862	634	607	176	4.63	22.7	270
1946	16.5	4.71	3.74	253	53.1	97.4	154	278	26.4	1.72	15.8	7.43	76.4
1947	59.3	27.8	74.4	25.4	9.1	355	1,049	169	816	23.5	0.94	39.0	220
1948	18.4	13.7	191	55.4	61.5	451	62.0	27.9	143	501	80.5	25.1	137
1949	3.71	12.1	10.0	206	406	242	119	257	353	80.5	22.4	87.0	148
1950	546	57.2	44.4	87.2	59.5	37.2	36.5	30.9	171	364	429	56.6	162
1951	12.0	7.87	6.52	7.36	18.1	45.3	137	231	713	1,616	75.4	593	290
1952	194	183	61.2	59.6	40.9	510	364	80.0	13.5	21.2	48.5	12.7	133
1953	2.81	5.04	4.71	4.17	6.77	32.5	84.7	105	7.44	3.81	2.45	2.59	21.9
1954	2.56	2.42	2.35	1.99	3.22	6.18	6.41	65.1	62.8	3.20	110	2.93	22.6
1955	146	8.03	4.50	42.1	162	31.7	18.4	281	125	206	36.8	34.5	91.2
1956	42.3	3.74	2.82	2.93	9.35	5.11	16.7	17.8	19.8	23.3	6.16	3.58	12.8
1957	2.83	3.54	3.84	3.52	4.12	4.36	13.3	107	349	94.0	7.59	79.4	55.8
1958	55.1	39.9	41.4	54.1	140	260	234	91.7	100	1,170	429	135	231
1959	34.7	25.7	17.5	24.6	64.4	171	263	105	30.6	24.9	13.5	12.1	65.4
1960	138	10.6	11.9	18.4	53.6	252	410	140	109	16.7	10.1	3.71	97.7
1961	53.4	22.4	48.1	15.2	104	375	509	548	81.8	253	119	1,304	285
1962	189	771	104	258	416	270	71.7	31.4	25.5	19.3	34.2	164	194
Mean of record	92.1	69.8	56.0	76.7	90.1	206	286	186	215	205	82.9	114	139

TABLE B.3. Streamflow record for flood of July 1965 on the Blue River

Time		Gage height	Discharge	Time		Gage height	Discharge
Day	Hour	(ft)	(cfs)	Day	Hour	(ft)	(cfs)
July 17	0200	5.39	47	July 20	0900	16.70	3,190
	0400	5.38	46		1000	18.20	3,720
	0600	5.48	58		1100	18.65	3,870
	0800	7.30	347		1200	18.84	3,940
	1000	10.30	1,140		1300	18.82	3,940
	1200	11.90	1,620		1400	18.70	3,900
	1400	11.40	1,470		1500	18.78	3,940
	1600	11.70	1,560		1600	18.92	3,970
	1800	10.10	1,080		1630	18.95	3,970
	2000	8.50	625		1700	18.94	3,970
	2200	7.70	425		1800	18.82	3,940
	2400	7.20	328		1900	18.60	3,870
July 18	0600	6.47	197		2000	18.25	3,720
	1200	6.17	148		2100	17.25	3,470
	1800	6.03	126		2200	16.40	3,090
	2400	5.89	106		2300	15.10	2,640
July 19	0300	5.84	99		2400	13.40	2,100
	0600	5.80	93	July 21	0200	10.50	1,230
	0900	5.80	93		0400	8.60	667
	1200	6.55	210		0600	7.95	493
	1500	7.38	366		0800	7.65	425
	1800	8.57	639		1000	7.45	385
	2100	7.52	385		1200	7.30	356
	2400	6.84	263		1400	7.17	328
July 20	0100	6.71	236		1600	7.03	309
	0200	6.60	219		1800	6.92	281
	0300	6.51	204		2000	6.82	263
	0400	6.46	195		2200	6.74	254
	0500	6.65	228		2400	6.65	236
	0600	7.55	395	July 22	1200	6.33	187
	0700	9.00	765		2400	6.15	158
	0800	11.20	1,410				

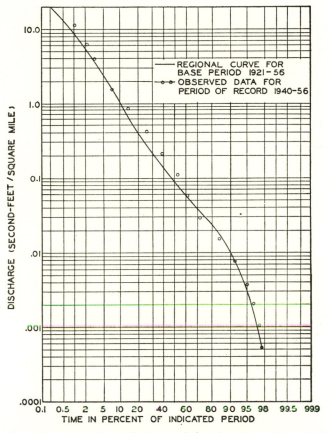

FIG. B.2.   Duration curves for daily flows of the Blue River. (From L. W. Furness, Kansas streamflow characteristics. 1. low duration, Kansas Water Resources Board, 1959)

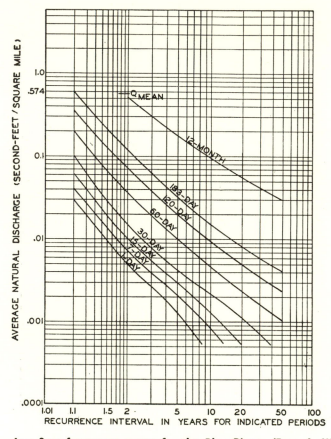

FIG. B.3. Low-flow frequency curves for the Blue River. (From L. W. Furness, Kansas streamflow characteristics. 2. Low-flow frequency curves, Kansas Water Resources Board, 1960)

APPENDIX C

LITTLE BLUE RIVER STREAMFLOW DATA

LITTLE BLUE RIVER

Location—Lat. 39°06'00'', long. 94°18'00'', in SW¼ SE¼ sec. 35, T.50 N, R.31W, at downstream side of right pier of upstream bridge on dual State Highway 78, 3 miles southwest of Lake City and 10½ miles upstream from mouth.

Drainage area—184 square miles. *Slope*—6.26 feet/mile.

Gage—Nonrecording prior to July 24, 1957; recording gage thereafter. Datum of gage is 719.15 feet above mean sea level, datum of 1929.

Stage-discharge relation—Defined by current-meter measurements.

Bank-full stage—18 feet.

Remarks—Base for partial-duration series, 2000 cfs.

LITTLE BLUE RIVER

AREA= 184 SQ MI

FIG. C.1.   The Little Blue River.

TABLE C.1. Discharge from the Little Blue River basin

Water year	Date	Peak discharge (cfs)	Mean annual discharge (cfs)
1948	May 20, 1948	6,000	
	July 26, 1948	3,200	
1949	Jan. 16, 1949	2,060	168
	Feb. 12, 1949	2,200	
	Mar. 31, 1949	2,060	
	May 22, 1949	2,800	
	July 12, 1949	2,080	
	Sept. 13, 1949	2,450	
1950	Oct. 22, 1949	5,580	101
1951	June 30, 1951	2,200	176
	July 6, 1951	2,060	
	July 11, 1951	6,400	
	Sept. 4, 1951	2,560	
1952	Oct. 6, 1951	2,060	127
	Mar. 10, 1952	3,690	
1953	Apr. 30, 1953	2,140	40.2
1954	Mar. 3, 1954	2,820	26.1
1955	May 29, 1955	4,000	59.5
1956	July 2, 1956	408	11.5
1957	July 1, 1957	1,680	22.8
1958	Aug. 1, 1958	4,350	131
1959	Apr. 28, 1959	1,290	45.6
1960	May 1, 1960	2,600	83.3
1961	Mar. 13, 1961	2,780	279.0
	Apr. 10, 1961	2,730	
	May 6, 1961	4,740	
	July 25, 1961	2,950	
	Sept. 4, 1961	2,220	
	Sept. T4, 1961	9,460	
	Sept. 25, 1961	4,100	
1962	Oct. 31, 1961	2,460	202.0
	Nov. 3, 1961	4,640	
1963	Oct. 13, 1962	1,900	43.7
1964	May 29, 1964	2,240	70.1
1965	June 5, 1965	2,240	162.0
	June 30, 1965	2,690	
	July 20, 1965	5,200	
	Sept. 22, 1965	3,500	
1966	July 9, 1966	2,360	119
	July 13, 1966	8,000	
1967	Apr. 2, 1967	2,240	185
	Apr. 14, 1967	2,000	
	June 13, 1967	3,250	
	June 22, 1967	5,410	
	June 25, 1967	2,370	
	June 29, 1967	3,390	
	July 27, 1967	2,180	

TABLE C.1. (continued)

Water year	Date	Peak discharge (cfs)	Mean annual discharge (cfs)
1968	June 1, 1968	2,150	150
	June 15, 1968	2,270	
	Aug. 9, 1968	2,040	
1969	June 23, 1969	3,250	257
	June 28, 1969	5,600	
1970	Oct. 13, 1969	5,700	264
	Apr. 19, 1970	4,570	
	May 16, 1970	2,100	
	June 13, 1970	3,640	
	Sept. 23, 1970	9,450	
1971	Jan. 4, 1971	3,500	107

TABLE C.2. Streamflow record for flood of July 1965 on the Little Blue River

Time Day	Hour	Gage height (ft)	Discharge (cfs)	Time Day	Hour	Gage height (ft)	Discharge (cfs)
July 19	0200	7.71	70	July 20	1130	25.03	5,200
	0400	7.95	85		1200	25.02	5,200
	0600	9.40	205		1800	24.90	5,100
	0800	12.10	523		2400	24.60	4,800
	1000	15.80	1,050	July 21	0600	24.15	4,430
	1200	18.50	1,620		1200	23.40	3,780
	1400	20.30	2,160		1800	21.85	2,840
	1600	20.90	2,400		2400	18.75	1,690
	1800	21.60	2,740	July 22	0600	15.20	959
	2000	23.40	3,780		1200	12.20	536
	2200	24.60	4,800		1800	10.70	345
	2400	24.92	5,100		2400	10.25	288
July 20	0600	24.98	5,200				

APPENDIX D

PRECIPITATION DATA, NORTHWESTERN MISSOURI, JULY 17-20, 1965

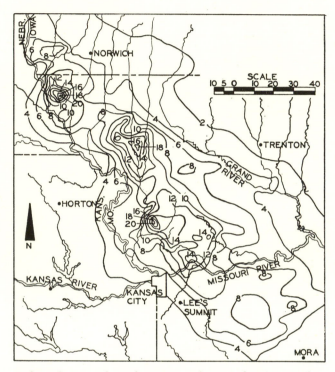

FIG. D.1. Isohyetal map of northwestern Missouri showing total precipitation for the storm of July 17–20, 1965. (From J. E. Bowie and E. E. Gann, Floods of July 18–23, 1965, in northwestern Missouri, Missouri Geological Survey and Water Resources, 1967)

TABLE D.1. Hourly rainfall, July 17–20, 1965

Hour	Kansas City airport				University of Missouri, Kansas City				Henrietta			
	17	18	19	20	17	18	19	20	17	18	19	20
1	...	...	0.04	...	...	...	...	...	...	...	0.06	0.02
2	...	...	0.10	...	...	...	0.1	0.1	...	...	0.10	0.05
3	...	...	0.04	0.04	...	...	0.1	...	...	...	0.01	0.06
4	...	...	0.10	0.69	...	...	0.1	...	...	...	0.12	0.11
5	0.54	...	0.11	0.07	...	...	0.1	0.3	...	...	0.35	0.09
6	0.01	...	0.69	0.05	0.6	...	...	0.4	...	0.04	0.50	0.09
7	...	0.06	0.04	0.17	0.1	...	0.2	0.1	...	0.01	0.42	0.04
8	...	0.08	0.02	0.06	...	...	...	0.1	...	0.17	0.31	0.04
9	...	0.01	...	...	...	...	0.1	0.1	...	0.08	0.22	...
10	...	...	0.79	...	...	0.1	...	...	...	0.01	0.38	0.02
11	...	...	0.20	...	...	...	0.4	...	...	0.02	0.23	0.01
12	...	...	0.25	...	...	...	0.3	...	...	0.01	0.22	...
1	...	...	0.04	...	...	...	...	...	...	0.01	0.25	...
2	...	...	0.10	...	...	...	...	...	...	...	0.10	...
3	...	...	0.16	...	...	...	0.1	...	...	...	0.15	...
4	...	...	...	...	...	...	0.2	...	...	...	0.18	...
5	...	...	...	...	...	...	0.1	...	...	...	0.02	...
6	...	...	...	...	...	...	...	...	...	...	0.04	...
7	...	...	0.03	...	...	...	...	...	...	...	1.83	...
8	...	...	...	...	...	...	...	...	...	...	0.07	...
9	...	...	...	...	...	...	...	...	...	...	0.08	...
10	...	...	...	...	...	...	...	...	...	...	0.18	...
11	...	...	...	...	...	...	...	...	...	...	0.04	...
12	...	...	0.01	...	...	...	...	...	...	0.03	0.03	...

Hour	Lee's Summit				Elm				Morse			
	17	18	19	20	17	18	19	20	17	18	19	20
1	...	...	0.01	...	...	...	...	0.25	...	...	...	...
2	...	...	0.06	0.01	...	...	0.03	0.04	...	...	...	...
3	...	...	0.02	0.01	...	...	0.11	0.03	...	...	...	...
4	...	...	0.01	0.18	...	...	0.75	0.08	...	...	...	0.3
5	0.04	...	0.01	0.11	...	...	0.18	0.05	1.5	...	...	0.3
6	0.05	...	0.05	0.11	0.03	...	0.06	0.05	0.2	...	0.1	0.6
7	...	0.05	0.33	0.18	0.02	0.02	0.06	0.05	0.1	...	...	0.3
8	...	0.05	0.01	0.11	...	0.45	0.03	0.13	...	...	...	0.1
9	...	0.04	0.01	0.04	...	0.08	0.06	0.11	...	...	...	0.1
10	...	0.01	0.18	0.01	...	...	0.04	0.01	...	...	0.2	...
11	...	...	0.29	...	...	...	0.52	...	...	...	0.1	...
12	...	...	0.02	...	...	...	0.10	...	...	...	...	0.1
1	...	...	0.16	...	...	...	0.31	...	...	...	...	...
2	...	...	0.04	...	...	...	0.08	...	...	...	...	...
3	...	...	0.29	...	...	...	0.44	...	...	...	0.1	...
4	...	...	0.09	...	...	...	0.46	...	...	...	...	...
5	...	...	0.17	...	...	...	0.29	...	...	...	...	...
6	...	...	0.04	...	...	...	0.20	...	...	...	...	...
7	...	...	1.54	...	...	...	0.15	...	...	...	...	...
8	...	...	0.01	...	...	...	2.40	...	...	...	...	...
9	...	...	...	...	...	...	0.11	...	...	...	...	...
10	...	...	0.01	...	...	...	0.28	...	...	...	...	...
11	...	...	0.01	...	...	...	0.17	...	...	...	...	...
12	...	...	0.03	...	...	...	0.14	...	...	...	...	...

INDEX

Aeration, zone, 152
American Geophysical Union, 13
American Society of Civil Engineers, 12
American Water Resources Association, 13
Annual
 flows, simulation, 140
 maximum flows, 32
 series, 34
 streamflows, maximum, 29
Aquifer
 confined, 156
 partially confined, 157
Area-depth curves, 77
Atmosphere, 51
 circulation, 53
 physical properties, 189
Average precipitation, 71

B

Bank storage, 100
Barnes, B. S., 116
Baseflow, 133
 curve, 101
 recession, 99, 102
Base time, 108
Beard, L. B., 34, 38
Benefit-cost analysis, 44
Benefits, incremental, 44
Benson, M. A., 38
Blue River, streamflow data, 194
Bureau of Reclamation, 79
Burkli-Ziegler equation, 113

C

Capillary action, 7
Carter, R. W., 122
Chicago, Ill., rainfall analysis, 61
Chow, V. T., 12, 31, 115, 116
Class A land pan, 171

Coefficient of storage, 162
Combinations, 19
Condensation, 5, 56
Consistency, establishment, 59
Continuity equation, 121
Convectional lifting, 56
Conversion factors, 183
 British to metric, 184
Coriolis force, 53
Costs
 annual, 42, 43, 46
 incremental, 44
Culvert size, selection, 45
Cusec, 14
Cyclone, extratropical, 57
Cyclonic lifting, 56

D

Dalrymple, T., 38
Dalton, J., 172
Dalton's law, 172
Damage,
 average annual, 44
 costs, 4
 expected, 44
Darcy's law, 153, 155
Data,
 climatological, 9
 collection of, 12
 general statistical, 183
 hydrologic, 9
 meteorologic, 9
 precipitation, 9, 64
 streamflow, 9
Davis, S. N., 161, 164
Dearlove, R. E., 149
Design discharge, 41
 event, 16
 parameters, 3

207

Deviation, standard, 25, 26, 28
DeWiest, R. J. M., 161, 164
Dewpoint, 188
Diffusion method, 171
Discharge, mean annual, 27
Distribution
  cumulative, 24, 27
  log-normal, 31
  normal, 26
Doldrums, 55
Dooge, J. C., 116
Double-mass analysis, 59
Drainage, 3
Drought, nonsequential, 141
Duration curves, 36, 199

E

Energy balance method, 171, 174
Environmental Science Services Adminis-
    tration, 9
Erosion control, 3
Evaporation, 5, 7
  amounts, 170
  computation, 172
  energy balance, 171, 174
  estimation, 171
  latent heat, 169
  pans, 171
  process, 169
Evapotranspiration, 169

F

Failure, risk, 23
Field capacity, 7, 8, 152
Fiering, M. B., 141
Fish, 3
Fitzgerald, T., 172
Flood
  control, 3
  damage, 42
  damage reduction, 3
  historical, 35
  plain management, 3, 4
  protection, 41
  reservoir, 5
  routing, 120
  wave velocity, 130
Flooding, levels, 4
Flow through porous media
  one-dimensional, 154
  two-dimensional, 156
Flow to a well
  nonequilibrium, 161
  steady, 159
  unsteady, 160
Foster, H. A., 31
Frequency curves, low-flow, 200
Front, 57

G

Gage, rain, 58
Godfrey, R. G., 122
Gould, B. W., 149
Graphic curve fitting, 32
Groundwater, 5, 8, 99, 152

H

Hail, 5
Harbaugh, T. E., 48
Harr, M. E., 164
Harris, R. A., 149
Hathaway, G. A., 112
Horse latitudes, 55
Horton, Robert, 86, 173
Hufschmidt, M. M., 141
Hurricanes, 58
Hydroelectric power, 37
Hydrographs, 96
  complex, 103
  outflow, 126
  streamflow, 8
  unit, 103
Hydrologic
  analysis, 4
  cycle, 5
  data, 9
Hydropower, 3

I

India, 15
Infiltration, 7, 85
  indices, 87
  process, 85
  rate, 8, 86
Initial abstraction, 88
Intensity-duration analysis, 59
  maps, 65–70
Interception, 5
International System of Units, 184
Irrigation, 3, 37
Isohyetal method, 72
Isohyets, 74

J

Jackson, B. B., 141

K

Keifer, C. J., 116
Kirpich, P. Z., 112

L

Lake Hefner, 173
Lake MacBride, evaporation study, 176
Lake Mead, 173

Land pan, Class A, 171
Lane, R. K., 179
Langbein, W. B., 35
Latent heat, 169
Levee, 4, 41
Little Blue River, 23, 28, 31, 34
   mass diagram, 139
   reservoir simulation, 138
   streamflow data, 201
Lloyd-Davies equation, 113
Logarithms, common, 192
Log Pearson Type III distribution, 29
Low flows, 36
   frequency curves, 142

M

Manning's equation, 112
Marais des Cygnes River
   low-flow frequency curves, 142
   storage requirements, 143
Mass diagram, 140
Mean annual flow, 146
Median, 26, 34
Meteorology, 51
Methods
   deterministic, 4
   probabilistic, 4
Meyer, A. F., 173
Meyers equation, 113
Missouri Geological Survey, 9
Mode, 26
Model
   conceptual, 7
   watershed, 7
Modeling, mathematical, 8
Moisture conditions
   antecedent, 93
   values, 153
Monsoon, 15, 56
Moran method, 149
Moran, P. A. P., 144
Morton, F. I., 173
Multipurpose project, 3
Municipal water supply, 37
Muskat, M., 164
Muskingum routing equation, 128

N

Navigation, 3, 37
Nonequilibrium flow to a well, 161
Nonsequential drought, 141

O

Optimal economic development, 44
Order number, 33, 34, 36
Orographic lifting, 56

P

Pan coefficient, 172
Partial-duration series, 35, 36, 42, 45, 64
Peak flow, equations for determination,
   113
Pearson, Karl, 29
Penman, H. L., 179
Permeability, 155
Permutations, 19
Phi index, 87
Plotting
   formula, 34
   positions, 33
Pollution control, 3
Polubarinova-Kochina, P. Ya., 164
Precipitation, 51, 56
   average annual, 15, 53
   consistency of record, 59
   data, 204
   duration of, 59
   estimating missing value, 58
   measurement of, 58
   over an area, 71
   at a point, 58
Pressure
   high, 54
   low, 54
Prism storage, 131
Probabilistic method, 144
Probability, 15, 17
   exceedance, 21
   in hydrology, 21
   paper, 29, 31
   reciprocal, 21
Probable maximum precipitation, 77

R

Rain, 5
Raindrops, 56
Rainfall
   annual maximum, 15
   excess, 7, 85, 88, 93
   intensity, 7, 9
Random numbers, 191
   variable, 15
Rapid City, S. Dak., 78
Rational method, 113
Record storms, 79, 80
Recreation, 3
Recurrence interval, 21
Reservoir
   operation, 137
   routing, 121
   size, required, 143
   storage, probability, 147
   yield, 137, 144, 149
Return period, 21, 64
Rippl diagram, 140
Rippl, W., 140

Risk, cost, 41
River basins, 10
Rohwer, C., 173
Runoff, 7, 96
    surface, 97
Russell, T., 173

S

Saturation, zone, 153
Scheidegger, A. E., 155
Second-foot-day, 14
Seddon, J. A., 130
Seepage, 7
Separation of base flow, 99
Simulation, 137
Skew, 25, 26
Snow, 5
Snyder, F. F., 108
Soil
    -cover complex, 88, 91
    groupings, 91
    storage capacity, 86
Soil Conservation Service, 88, 109
Springs, origin, 51
Stall, J. B., 144
Station-average method, 71
Statistical methods, 15
Storage
    bank, 100
    depression, 7, 88
    detention, 7
Storm sequence, 76
Storms, transposition, 79
Streamflow, 96
    annual maximum, 29
    data, Blue River, 194
    data, Little Blue River, 201
    routing, 126
Stream rises, classification, 8
Surface runoff, 8
Synthesizing, 4
Synthetic storms, construction, 74

T

Talbot equation, 113
Theis, C. V., 161
Theissen, A. H., 72
Theissen polygon, 72
Tholin, A. L., 116
Thunderstorms, 56
Time of concentration, 98, 110, 111, 113
Transmissibility, coefficient, 162

Transpiration, 7, 169
Triangular unit hydrograph, 109
Typhoons, 58

U

Unit hydrograph
    application, 106
    construction, 105
    limitations, 112
    Snyder, 108
    synthetic, 107
    triangular, 109
U.S. Geological Survey, 9

V

Valley storage, 130

W

Walton, W. C., 155, 161, 164
Water
    conservation, 3
    physical properties, 190
    quality improvement, 3
    supply, 3
Water Resources Research, 12
Water Supply Papers, 9
Watershed
    management, 3
    response, 8
Weather, 54
    orographic effects, 56
    seasoned effects, 56
Wedge storage, 131
Well function, 162
Wenzel, L. K., 163
White, J. B., 149
Wildlife, 3
Wilting point, 152
Wind directions, 54
    prevailing, 55
Wycoff, R. L., 48

Y

Yuan, P. T., 31

Z

Zone
    groundwater, 7
    root, 7
    saturation, 7